A HISTORY OF
TRAVEL IN

50

VEHICLES

A **HISTORY** OF **TRAVEL** IN

50

VEHICLES

PAULA GREY

Introduction by **PHILLIP HOOSE**

Tilbury House Publishers
12 Starr Street, Thomaston, Maine 04861
800-582-1899 • www.tilburyhouse.com

First hardcover edition: August 2016
ISBN 978-0-88448-399-1

Library of Congress Cataloging-in-Publication Data

Names: Grey, Paula.
Title: A history of travel in 50 vehicles / Paula Grey.
Other titles: History of travel in fifty vehicles
Description: First edition. | Thomaston, Maine : Tilbury House Publishers,
    2016. | Series: History in 50 | Includes bibliographical references and
    index.
Identifiers: LCCN 2016016369 (print) | LCCN 2016017688 (ebook) | ISBN
    9780884483991 (hardcover) | ISBN 9780884484912 (pbk.) | ISBN 9780884483984
    (ebook)
Subjects: LCSH: Vehicles--History. | Vehicles--Technological
    innovations--History. | Travel--History. | Transportation--History. |
    Inventions--History.
Classification: LCC TA1015 .G74 2016 (print) | LCC TA1015 (ebook) | DDC
    629.04/609--dc23
LC record available at https://lccn.loc.gov/2016016369

Text designed by Jonathan Friedman, Frame25 Productions
Cover designed by John Barnett, 4 Eyes Design

Printed in China through Four Colour Print Group, Louisville, Kentucky

15 16 17 18 19 20 4CM 5 4 3 2 1

Table of Contents

Presenting the History in 50 Series

by Phillip Hoose

The *History in 50* series explores history by telling thematically linked stories. Each book in this series includes 50 illustrated narrative accounts of people and events—some well-known, others often overlooked—that, together, build a rich connect-the-dots mosaic and challenge conventional assumptions about how history unfolds. In *A History of Civilization in 50 Disasters,* for example, Gale Eaton weaves tales of the disasters that happen when civilization and nature collide. Volcanoes, fires, floods, and pandemics have devastated humanity for thousands of years, and human improvements such as molasses holding tanks, insecticides, and deep-water oil rigs have created new, unforeseen hazards—yet civilization has advanced not just in spite of these disasters, but in part because of them.

History in 50 is a canny, fun, and logical way to present history. The stories are brief, lively, and richly detailed. They work as narrative and also as bait to lure readers to well-selected source material. History in these books is not a stuffy parade of generals, tycoons, and industrialists, but rather a collection of brief, heart-pounding non-fiction narratives in which genuine calamities overtake us, genuine athletes leap skyward on feet of clay, and genuine discoverers labor bleary-eyed through the night to take us to the depths of the ocean, explore the vast reaches of space, unlock the genetic code, or develop a vaccine that saves millions of lives. It's history that bellows and shivers and roars.

And who doesn't love lists? Making a list of fifty great episodes of any kind invites—*demands*—debate. Even if the events aren't ranked (they're in chronological sequence), something always gets left out. I just finished reading A *History of Civilization in 50 Disasters*. I paged wide-eyed through plagues, eruptions, famines, microbes, and vaccines that worked or didn't. From my reading chair I took on dust storms, melt-downs, and epidemics at all scales that claimed my unwavering attention. When I closed the book and looked up, blinking, my first thought was, "Unbelievable. How have we ever made it through all this?"

But these feelings quickly gave way to a surge of indignation: *Where was the Tri-State Tornado of March 1925?* It's my favorite disaster—one that hit home. Actually a series of twisters, the Tri-State storm ripped through Missouri and Illinois before closing in on my great-grand-parents in southwestern Indiana. Seven hundred people were killed in what is commonly ranked as the worst tornado ever. Contemporary meteorologists agreed that it was surely a category five twister, and yet it didn't make the top fifty disasters? I needed to lodge a protest.

But then I realized that my pique was a good thing. The book had made me care. The stories had swept over me and shaken my certainty like the 1906 San Francisco earthquake. And I realized that many read-ers will have the very same reaction: *Hey, where's my favorite episode?* It will spur debate. I imagine smart teachers asking students to describe their own favorite historical episodes, backing up findings with research. I imagine readers of all ages heading back to their bookshelves to support their arguments.

History is rewarding, but in my experience most people have to be led to it. So-called Reluctant Readers are mainly reluctant to be bored. They require, and deserve, historical material that meets them partway. History with menacing characters, even if some of them are invisible (germs); history replete with tough decisions; crisp episodes that leave you wondering what you would have done in that situation; history moved by people just like us, often from the humblest of origins, struggling in their daily lives while reaching for greatness—that's the history that works for most readers. And that is the history we have in this brilliant new series. The writing is clear and exciting, punchy stories that are, on average, two pages long. I have high hopes for the *History in Fifty* series, and it gives me pleasure to enthusiastically endorse it. Why? Because it works.

PHILLIP HOOSE is the National Book Award–winning author of *Claudette Colvin: Twice Toward Justice* and *The Boys Who Challenged Hitler: Knud Pedersen and the Churchill Club*.

Introduction

Vehicles and Travel

French author Jules Verne published his popular novel *Around the World in Eighty Days* in 1873. The book charts the journey of Phileas Fogg, a wealthy Londoner, and his French valet, Passepartout, as they attempt to win a bet by completing a trip around the world in what was then the astonishingly short time of eighty days. (A valet was a manservant, and Passepartout translates from French into English as "goes everywhere.") The bet is a large one that Fogg can ill afford to lose, and his friends believe it will be impossible to win. But Fogg has mapped out the journey and is convinced that he can prevail. His not-so-secret weapons? The latest in speedy modern travel: steamships and steam locomotives. Verne's book highlighted a new era in travel, with links between major cities worldwide and vehicles that ran on reliable schedules.

Phileas Fogg's journey wasn't smooth and easy; what kind of an adventure book would that be? Bad weather, a collapsing railroad bridge, and miscommunication with his valet led to delays and adventures. Where ships and trains weren't available, Fogg had to rely on more traditional transportation, including an elephant in India and a wind-powered sledge across the U.S. prairie.

Verne's novel highlighted the astonishing speed and ease of travel in 1870. For the first time in history, people could reach almost any part of the world in relative comfort on a predictable schedule. What a remarkable advance! And how amazed Verne and his readers would have been if someone had told them that a little over 130 years later, a man would *fly* around the globe in 67 hours.[1]

Australian National Maritime Museum

A tall ship, a steam ship, and a biplane meet in Sydney Harbor, Australia, in 1935.

The history of travel is the story of humankind's constant search for new and better ways to move about in the world.

Nobody knows why the first humans left their birthplace in Africa. Suggested motives include massive droughts, changes in climate that lowered sea levels to expose land bridges, or the need to follow the migrations of prey animals. Maybe competition among families or tribes forced people to seek new territories. Or maybe some of our prehistoric ancestors were curious about what marvels lay beyond that next ridge. Regardless, almost from the beginning, humans have been a traveling species. And an imaginative one.

Picture our prehistoric ancestors moving northward through Africa. Limited by their least fit clan members, their rate of progress must have been no more than 2 miles (3 kilometers) per hour even in the best conditions. What would they have thought of John Adams, who is said to have traveled some 30,000 miles (50,000 km) in his life at a time when most Americans never left the county in which they were born? Adams's plodding horses and cumbersome ships, and his average speed of 4 miles (6–7 km) per hour (a century before Phileas Fogg), would have amazed his prehistoric ancestors no less than a jetliner flight would have amazed him.

How we travel has changed over time, but the reasons remain the same. Every day, people ride subways, buses, trains, automobiles, rickshaws, pedicabs, horses, and mules to get to and from work or run errands. Refugees seeking safety walk miles or cram into leaky boats to cross unknown seas. Far-flung family members gather to celebrate holidays and important events. People with time and money to spare seek out new places and experiences for education and relaxation. And we are still just beginning to explore the depths of the oceans and the far reaches of space.

This book looks at how travel has changed over the ages. Its focus is vehicles built for *travel*—the movement of people—as opposed to *transportation*—the movement of goods. Vehicles used for travel sometimes carry goods as well, but their primary purpose is to take us where we want or need to go.

A complete list of the vehicles invented for travel through the ages would be much, much longer than the 50 presented here. Some of the vehicles in this book were selected because they are historically significant; some were first built centuries ago and have evolved into forms still in use today; some represent a particular geographic region or culture; and some are simply interesting and unusual.

In many cases, it's difficult to know who designed a specific vehicle, or when. Sometimes the best evidence of early travel is an image on a cave wall or a piece of pottery. Most ancient vehicles made of leather, wood, reeds, and other biodegradable materials decompose long before archaeologists go digging for them. Horses were domesticated and boats were built by cultures around the world making independent advances. And many of the vehicles we use today are the work of multiple inventors building on each other's successes over tens of years. An estimated

10,000 patented inventions were required to create the modern automobile. For all these reasons, many dates in this book are approximations based on archaeological evidence or the point at which a design or invention first resembled the vehicle we know today.

The same type of vehicle might develop differently in different cultures, with variations for local conditions. Numerous civilizations that grew up along major waterways built boats. Where large trees were common, people built dugout canoes, but where trees were scarce, people made rafts and boats from reeds. People in Europe and Asia domesticated horses, but horses became extinct in North and South America at the end of the last Ice Age and were reintroduced from Europe in the sixteenth century. And not every culture adopted the wheel. The Inca and other South American tribes found that a hand-pulled travois worked better on grasslands, sandy soil, or steep hills than a wheeled cart, while people in snowbound northern regions traveled on skis or by dogsled.

New vehicles have been invented throughout history, but *reinvention* and adaptations of existing concepts are equally important in the story of travel. Early makers of skis and sails wouldn't recognize today's versions made from lightweight carbon fiber and synthetic materials, much less the concept of using a solar sail to power a spacecraft. Automobiles, trains, buses, and ships have all made the move from steam engines to internal combustion powered by gasoline. Subways moved from steam to electric power. Advances in aviation technology during World War II led to lighter, more powerful engines that were exactly what was needed to make motorcycles more workable. And the invention of the turboshaft jet engine was critical to the success of modern helicopters.

New vehicles called for new infrastructure to support them, and infrastructure, in turn, spurred adoption of the vehicles. Horse-drawn carriages needed roads, and automobiles need even stronger roads. Trains need rails. Traffic signals and air traffic control towers evolved because vehicles made them necessary. Gas stations popped up along highways as cars started making longer journeys and needed to refuel. Once elevators became practical, cities began to grow *up* rather than *out.* Canada, Alaska, and the Soviet Union all built railways to remote areas for access to their natural resources; soon people were building new settlements along the rail lines.

Another factor in the widespread adoption of vehicles is economics. The first electric-powered carriage was introduced in the 1830s, and many early cars and buses were powered by electric engines. But by 1935, electric vehicles had all but disappeared due to improvements in gas-powered engines, demand for cars that could make longer trips, and the lower price of gas-powered cars. Now, given advances in battery technology and concerns about the impact on climate change of burning fossil fuels, electric cars are poised for a comeback. The destruction of railway lines in Europe during World War II required massive post-war rebuilding, and improvements in electric locomotive technology made electricity an attractive and affordable option for new railways. But the U.S., with no need to rebuild, continued to rely on existing diesel technology.

Political and social concerns have also played a role in the development and use of vehicles. Resistance to the loud sonic booms created by supersonic jets led to the demise of that form of travel. Many countries have banned the use of human-pulled rickshaws as oppressive and degrading to the rickshaw "driver." Subways were once viewed as possibly satanic because they existed in the realm of the Underworld, and people worried that digging so deep would release deadly germs or cause roads and buildings above the tunnels to cave in. The first automobiles scared pedestrians and horses. Carriage makers and hackney drivers saw autos as a threat to their jobs and wanted them banned.

Some fears were very real. Early trains ran on steam engines, which could explode, killing engineers and passengers. Dirigibles (Chapter 29), powered by highly flammable hydrogen, were even more dangerous. Airplanes could and did crash, and ships sank. But inventors continued to find solutions to these problems, and people continued to travel.

Early immigrants to America and pioneers heading west in covered wagons left home with the knowledge that they would probably never return. Today, people can fly almost anywhere in a day. Increasing travel has led to globalization of goods and services. Travelers can get sushi in New York and a McDonald's hamburger in Tokyo. Travel exposes people to new cultures, foods, traditions, and beliefs, but global travel also makes it easier for infectious diseases, such as Ebola, and invasive species of plants and animals to migrate to new environments. Optimists say increasing travel will lead to better global communication and understanding. Pessimists say it will result in countries losing their unique cultures and becoming all the same. Both may be at least partly right.

Where will humans travel next? How will we get there? The history of travel makes one thing clear: we'll go wherever curiosity takes us, inventing and reinventing the vehicles we need along the way. Like Phileas Fogg and Passepartout, there will always be people eager to take advantage of new advances, and the exciting new vehicles of today will fade into history just as Fogg's steamships and steam locomotives have done.

Shoes
Feet First!

Walking has always been humans' primary means of travel. Early humans migrated from Africa to India, Asia, and eventually Europe and the Americas largely on foot. On July 20, 1969, astronaut Neil Armstrong took the next step in this migration by walking on the surface of the moon.[1] Even today, the primary vehicle for millions of people worldwide is their own two feet.

Prehistoric foot travel had big drawbacks. It was slow, and travelers could take only as much food, water, and household goods as they could carry. Small children had to be carried. Prior to the development of shoes, hot sand, snow, ice, sharp rocks, and thorny bushes made trekking difficult and painful. Yet early peoples traveled thousands of miles on foot to settle new parts of the globe.

Based on fossil evidence and genetic (DNA) analysis, scientists believe that modern humans (*Homo sapiens sapiens*) developed in Africa some 200,000 years ago. Approximately 60,000 to 80,000 years ago, they traveled by foot to Asia, then settled what are now Indonesia, Papua New Guinea, and Australia. By 35,000 years ago, modern humans were firmly established in most of Europe. Finally, approximately 20,000 years later, humans trekked from Asia to North, Central, and South America—again on foot, crossing the Bering Strait land bridge.[2]

The earliest shoes archaeologists have found are sandals dating from approximately 7000 or 8000 BC in the Fort Rock Cave in Oregon.[3]

The oldest leather shoe, a piece of cowhide laced with a leather cord, was found in a cave in Armenia in 2008 and is believed to date back to 3500 BC. The terrain around the cave was rugged, with sharp stones and prickly bushes. Temperatures in the area could range from up to 113° Fahrenheit (45° Celsius) in summer to below freezing in winter.

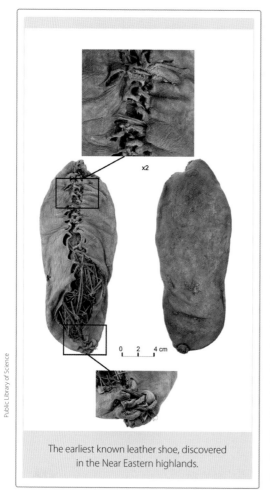

x2

0 2 4 cm

The earliest known leather shoe, discovered in the Near Eastern highlands.

Public Library of Science

Shoes would have enabled people in the region to cope with the terrain and the temperatures. Obsidian, an igneous rock from at least 75 miles (120 km) away, was found in the same cave, showing that they traveled long distances.[4]

Despite the relatively recent age of the early shoes found to date, scientists believe that humans were wearing shoes as much as 40,000 years ago. Shoes made of reeds, papyrus, or even leather decompose easily, leaving few or no traces. But people who don't wear shoes have wider feet and bigger gaps between their big toe and the other four. In studying bones of the smaller toes of fossilized skeletons, scientists observed that the thickness of these bones decreased somewhere between 40,000 and 26,000 years ago. They believe the change was the result of wearing shoes.[5]

The materials used to make shoes varied with climate and region. Ancient Egyptians made sandals from papyrus and palm leaves. The

Masai of Africa used rawhide; the people of India used wood; and the Chinese and Japanese used rice straw. South Americans wove sandals from the leaves of the sisal plant, and the Anasazi people of Mexico and Arizona used the yucca plant.[6] The Dutch developed wooden clogs. People living in arctic climates created snowshoes—a hardwood frame with rawhide lacings—which distribute a person's weight over a larger area so that the foot doesn't sink into the snow.

Today, shoe manufacturers use rubber, plastic, cloth, and other materials in addition to leather. Most soles are made from synthetic materials such as ethylene, vinyl acetate, rubber, and polyurethane, which provide better traction, durability, and water resistance than leather soles.[7] Special shoes are available for running, bowling, and other sports. And shoes are often considered an important fashion accessory.

Snowshoeing for sport.

Thinkstock/Ingram Publishing

Walking as Sport

Beginning in the late seventeenth century, *pedestrianism* (walking) became a popular spectator sport in the British Isles. By the 1800s, the sport had spread to the U.S. and Australia. Some towns built indoor tracks. In one variation, athletes competed to see who could walk a long distance in the shortest time. Captain Robert Barclay Allardice, known as the "Celebrated Pedestrian," earned fame by walking one mile every hour for 1,000 hours, day and night, from June 1 to July 12, 1809, napping and eating between his hourly jaunts. Presumably the wagers on his attempt—the equivalent of $8 million today—kept him moving. Ada Anderson later bettered Allardice's record by walking a quarter mile in each 15-minute period over 1,000 hours,[8] and in 1867, Edward Payson, a reporter for the *New York Herald*, won a $10,000 prize by walking from Portland, Maine, to Chicago, Illinois, a distance of 1,136 miles (1,832 km), in 30 days.

Pedestrianism later evolved into the amateur sport known as racewalking, which became an Olympic event at the 1908 Olympic Games in London.[9] The 2016 Summer Olympics in Rio de Janeiro, Brazil, will include a 20-kilometer (12.4-mile) racewalk for men and women.

Moon Boots

NASA

Buzz Aldrin booted up for his moon walk, July 20, 1969.

You would think that the boots Neil Armstrong wore for his moon walk would be carefully preserved in the Smithsonian's Air & Space Museum. But scientists were concerned about what might come back to earth in the dust on those boots. Some strange organism or dangerous disease? There was no way to know. To be safe, Armstrong's boots were left in space.[10]

Litters
Moving Those Who Cannot Walk

The Atsina were a nomadic Arapaho tribe that followed the bison herds across the American plains. That meant moving camp frequently, as here, with belongings carried on travois behind the horses. By 1888 the Atsina were confined to a reservation at Fort Belknap, Montana.

Edward S. Curtis Collection, Library of Congress

Infants, small children, the elderly, and the sick and injured were unable to travel on foot and had to be carried. Carrying an infant is easy; a simple sling over a mother's shoulder will do the trick. But for one adult to carry another, even a short distance, requires strength and endurance. Even a small person gets heavy quickly.

Early humans solved the problem of carrying heavy loads by developing a wheel-less vehicle called a travois. It consisted of a platform

Painting of a medieval litter en route to Constantinople.

or netting mounted on two long poles that were lashed together in a V shape with the pointed end at the front. Items were loaded on the platform, and a person or animal pulled the travois.

A litter—a sling or platform carried between two poles like the stretchers used to remove wounded soldiers from a battlefield—is similar to a travois, but the poles are parallel to one another. A litter is carried on the shoulders of two or more people rather than being dragged.

Litters developed in most early civilizations, including ancient Rome, Persia, Vietnam, England, Spain, France, India, Ghana, China, Portugal, Korea, Japan, Turkey, and Germany. They appear in Egyptian paintings and are mentioned in the Book of Isaiah in the Old Testament.[1] While litters might look slightly different from one culture to another, they all took the form of a chair or platform carried by two or more people.

Smaller litters could be open or enclosed in curtains for protection from the weather. Larger litters, such as those used by Chinese emperors, resembled small rooms and required up to a dozen men to carry them. It has been said that King Henry VIII of England was forced to add extra carriers as he became increasingly obese toward the end of his reign. In some cultures, litters were only for royalty, the wealthy, or politically important people, while the average citizen had to walk.

In England, sedan chairs (the British name for litters) were a major mode of transportation throughout the seventeenth and eighteenth centuries and the early part of the nineteenth century. They were especially useful in London and other cities. Narrow streets and tightly

In this late nineteenth-century oil painting by S. Borgelli, a lady in a sedan chair is greeted by a gentleman while the chairmen take a (presumably) much-needed rest.

packed houses did not always leave room for a horse and carriage, and it was often faster to walk. In an era of filthy, muddy streets, riding in a sedan chair let people stay clean and dry.[2] In many cities, sedan chairs were licensed and available for hire, like taxicabs today.

Because sedan chairs remain horizontal while being carried up and down stairs, the smaller models could be taken into people's houses. People could travel from one location to another while remaining fully sheltered from bad weather. In Bath, England, where invalids went for healing sessions in hot mineral baths, sick people could climb from their beds directly into a sedan chair. They could then be carried to the baths and back again, avoiding contact with the chilly outside air.[3] Sedan chairs also made it possible for people to travel without being seen. Those on the run from debt collectors or the police could stay out of sight, and politicians could conduct secret meetings.[4]

In most parts of the world, people stopped using litters when wheeled vehicles and horses or other animals became more common and road quality improved. Litters are still used today, however. In Chinese mountain resorts, such as the Huangshan Mountains, tourists ride along scenic paths to reach spectacular views in a simple cane chair on two stout bamboo poles.[5] The annual Sedan Chair Race in Hong Kong, a major event, has raised over $63.3 million for approximately thirty Hong Kong charities since 1975.[6]

Dugouts
The First Boats

Early civilizations were centered in river valleys, where water was plentiful, the soil was fertile due to periodic flooding, and wild animals came to drink, making hunting easier. Edible plants grew along the riverbanks, and fish and shellfish provided an additional source of food. When humans began to travel, they traveled by water as well as on land.

The oldest known wooden boat is the Pesse canoe, found in what is now the Netherlands. The canoe was constructed from a hollowed-out tree trunk somewhere between 8200 and 7600 BC.[1]

Making a dugout was a long and arduous process. The wood had to be fresh, not rotted, ruling out many fallen trees. Given the size and weight of a log big enough to create the desired boat, builders preferred logs that were close to the water. Frequently, the best log was a living tree. Archaeologists have demonstrated that it is possible to cut down a tree with a stone ax, although the process could take several days.[2]

Once the tree was felled, the next step was to strip the bark. The log was then hollowed out by using controlled fire to scorch a layer of wood for removal with a stone adze, repeating many times until the job was done. An alternative method was to chop parallel notches into the wood, then split out and remove the wood between the notches. Boat makers needed to remove enough wood to make the dugout light and buoyant while leaving enough thickness for strength and durability.

Dugout canoes developed all over the world. The Dufuna canoe from Nigeria is the third oldest found worldwide.[3] Many prehistoric dugouts have been found in Scandinavia, and these boats are thought to have been common among the Stone Age people of northern Europe. The Māori traveled from eastern Polynesia in big dugouts, called *waka*, to settle New Zealand.[4] Roman records have been found documenting how Slavs in Eastern Europe built large dugouts and sold them to the Vikings for use in naval battles in the ninth and tenth centuries. Meanwhile, in North America, the indigenous people of the Pacific Northwest were creating beautifully carved totem poles and dugout canoes.

Dugout canoes in the Okavango Delta region of Botswana. Are wet riverside grasses packed into idle canoes to prevent the wood from splitting in the hot sun?

Teinesavaii

A dugout fishing canoe with outrigger, or *ama*, on Savai'i Island, Samoa.

While the word *canoe* makes us think of a two-person boat, the dugouts used by the Vikings to attack Constantinople could hold between 40 and 70 warriors, and the vessels used by the Māori were equally large. Outriggers—long, thin hulls that were attached in parallel with the main hull—kept large canoes stable, particularly in rough ocean waters.

Dugouts are still in use today in some parts of the world. Tourists visiting the Mergui Archipelago in Myanmar can watch the traditional inhabitants, the Moken, making and using dugout canoes as they have for tens of thousands of years.[5] In the Pacific, Solomon Islanders still use dugouts to travel among islands. And native peoples as diverse as the Cherokee Indians in North America, the Haida of the Pacific Northwest, and Australian Aborigines continue to pass down the craft of making dugout canoes as part of their cultural heritage.

Reed Rafts and Boats
As Old as Dugouts?

Egyptians living in the Nile River delta had few roads but numerous waterways. The easiest way to travel was often by water.[1] Lacking access to trees large and strong enough to be used for boatbuilding, the Egyptians relied on a plant that grew in abundance—the papyrus reed—to construct rafts and boats. Other civilizations throughout the world did the same with reeds native to their regions. Reed boats and illustrations of them have been found throughout the Mediterranean and along the Atlantic coasts of Spain and Morocco.[2] Inhabitants of Peru, Bolivia, and Easter Island built rafts and boats from totora reeds. Native Americans around San Francisco Bay used tule reeds,[3] and East Asians used bamboo.[4]

The oldest discovered remains of a reed boat, believed to be 7,000 years old, were found in Kuwait,[5] but humans may have used reed rafts far earlier. Reeds are a natural fiber that decomposes easily, leaving few traces for archaeologists to uncover. The ancient reed boats that have been found were preserved in part by the tar used to waterproof them.[6] Reed boats are depicted in early petroglyphs (rock engravings), providing another way to determine when and where they were used.

Reed rafts were easy to construct and did not require the specialized tools and skills needed to build wooden boats.[7] Reeds were tied into bundles, and the bundles were lashed together with vines or some

Kon-Tiki Museum

The reed boat *Ra II* crossing the Atlantic in 1970.

other tough fiber. The hollow reeds floated easily, and water splashing up on the raft quickly drained through the small gaps between bundles. Because the rafts floated flat on the water, they were stable and difficult to capsize.[8] Small rafts were used for short coastal or river voyages; larger rafts could venture offshore and carry more passengers and animals or larger cargoes of trade goods.

Reed boats were constructed much like rafts, but with rounded, sickle-shaped hulls, raised sides, and raised prows in the front and rear. Some boats also had deckhouses and masts for sails.[9] An exterior coating of tar kept water out. Reed boats are still built and used in Lake Titicaca in the South American Andes and Lake Chad in Central Africa.[10]

Papyrus

A papyrus plant growing in water

Ancient Egyptians used the papyrus plant for everything from cooking to writing. The tough, fibrous parts of the plant could be woven into mats, baskets, or sandals; tied in bundles to create reed boats; or burned as fuel. The lower, softer part of the plant could be cooked and eaten. But the use that most people have heard of is the creation of a kind of writing paper.

Papyrus was used as a writing material as early as the fourth millennium BC. Pieces of the wet pith (tissue in the plant's stems) were laid flat, side by side. A second layer of pith was added at right angles to the first, and the two layers were pressed together to fuse them, then smoothed with a stone or shell and dried in the sun. The labor required for its creation ensured that papyrus was expensive and not normally used by the average person, but saved for official correspondence.

In 1969, Thor Heyerdahl, a Norwegian explorer, set out to prove that a papyrus-reed boat could have sailed across the Atlantic Ocean. His first boat, the *Ra*, was built of 12 tons of papyrus and carried a crew of seven. The *Ra* set sail from Safi, Morocco, and traveled 2,700 nautical miles (5,000 km) before storms and structural failures ended the

Auntjojo

A family sculling a reed boat off Uros Island on Lake Titicaca, Peru.

trip. Undaunted, Heyerdahl set sail ten months later in the *Ra II* and succeeded in traveling 3,270 nautical miles (6,100 km) in 57 days, crossing the Atlantic from Safi to Barbados. Heyerdahl had demonstrated that ancient boatbuilders could, indeed, compete with modern marine technology.[11]

Planked Wooden Boats
The Shipbuilder's Art

Some archaeologists think the first wooden boats were dugout canoes with planks added to their sides to hold more cargo.[1] Others point out that the first wooden ships built by the Egyptians and Phoenicians imitated the shape of earlier papyrus boats, including the high, curved bow and stern, which weren't needed when working with wood. These experts believe that our ancestors moved from reed boats to wooden boats when copper and iron tools to build them became available, sometime around the middle of the fourth millennium BC.[2]

Wood has big advantages over papyrus and other reeds. As reeds become saturated with water, they can lose their shape and sink or pull apart. One estimate is that the working life of an ancient reed boat was approximately one year.[3] Wooden boats are stronger and have a much longer useful life.

The Khufu ship, found in an Egyptian tomb, is one of the best examples of an early wooden boat. The ship had a flat bottom with sides built of wood planks that were joined by mortises and tenons and "sewn" together with approximately 16,400 feet (5,000 meters) of cordage. Once the boat was in the water, the wood swelled, sealing any small cracks.

Joining planks with tough vines, hemp rope, strips of leather, or some other cordage gave a boat additional flexibility, and a more flexible boat was less likely to be damaged in a collision or when run up on a beach to load or unload goods. It was easy to replace one or more planks or take the entire boat apart, transport it overland, and reassemble it where needed.[4]

The primary method of propulsion for early wooden boats, called galleys, was rowing. The largest of these were called *penteconters*, Greek for *fifty-oared*, because they carried 50 oarsmen—25 per side. Adding more oarsmen made a boat capable of higher speeds, but also required that it be longer and heavier, making it harder to maneuver. The solution was to add a second, upper deck and group the rowers into two tiers per side; these vessels came to be called *biremes*. Eventually a third tier was added to create the *trireme*.

The Khufu ship was sealed into a pit at the foot of the Great Pyramid of Giza, in Egypt, around 2500 BC. Planked with Lebanon cedar, the slender vessel is 143 feet (43.6 meters) long. After a restoration that took many years, it is displayed in the Khufu Boat Museum near the Great Pyramid.

Berthold Werner

Triremes could ship a total of 170 oarsmen: 27 per side for the first two levels and 31 per side on the top level. The rowing positions were staggered so that each rower had room to work without tangling oars with the rowers above or below him, and the top-tier oars were deployed farther outboard than the tiers beneath them, through an overhanging deck.[5] Still, it was difficult to coordinate the strokes of the three tiers for maximum effectiveness. The demands of strength and endurance were greatest on the oarsmen of the top tier, but the bottom-tier rowers suffered the greatest discomfort from the rowers close overhead and the water shipped through their open rowing ports.

At 120 feet (37 meters) long with a speed of 6 to 8 knots and the ability to turn 90 degrees in a ship's length, the trireme was faster and more maneuverable than other warships of its time. Triremes are credited with helping the Greeks defeat the much larger invading Persian navy at the famous battle of Salamis in 480 BC and with the rise of Athens as a major power in the Aegean Sea.[6]

Galleys weren't totally dependent on oarsmen; many had sails as well. Rowers were used close to shore or when a ship needed to maneuver carefully or make progress into the wind, but sails were better for open water and long distances when the wind was favorable, especially on ships with wider, deeper hulls for carrying large quantities of trade goods. Galleys were used in warfare to ferry troops from one place to another, and soldiers from one ship could board another, capturing the crew and the vessel.

Reconstructed Greek biremes. Note the two tiers of oars on each ship.

Deutsches Museum

The development of the ram around the eighth century BC added a new dimension. A ram was a large projection, usually covered with metal, mounted on the bow of a ship below the waterline. It could punch a hole in the hull of an enemy ship, causing it to sink. A larger, heavier ship could put more force behind its ram and do more damage, but a smaller, faster ship could outmaneuver an attacking ram. Shipbuilders began to make trade-offs in ship design, depending on the purpose of the vessel.

Prisoners of war and slaves were used to man a war galley's oars when not enough free citizens volunteered. Galleys were still in use until the reign of King Louis XIV of France in the early 1700s as convict ships for felons.[7] Prisoners were branded with the letters G-A-L and forced to serve at least 10 years under harsh conditions. Many did not survive.[8]

Mortise-and-Tenon Construction

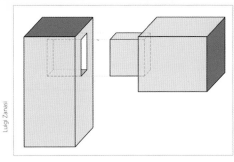

Luigi Zanasi

Diagram of a mortise-and-tenon joint.

Mortise-and-tenon construction is still used today. The tenon, or peg, carved on one piece of wood is inserted into a mortise hole in another piece. A mating tenon and mortise are cut to fit exactly so that the two pieces join with no gap between them.

Naval Warfare

Codex Skylitzes Matritensis Biblioteca Nacional de Madrid

This image from an illuminated manuscript of the twelfth century shows Roman sailors deploying Greek fire against a rebel.

Ancient warships had a number of weapons besides the ram. The upper decks of triremes provided shelter for archers and spearsmen. Larger ships could mount a catapult and hurl large stones and pots of poisonous vipers and scorpions.[9] But the most frightening weapon of all was Greek Fire.

Greek Fire was an ancient form of flamethrower—a jet of flame sprayed from a tube mounted on the prow of a ship. Although no one is certain what exactly was used to create Greek Fire, it appears to have been a petroleum-based liquid. Accounts of ancient battles indicate that the fire floated on the water and was almost impossible to put out. For sailors on wooden boats, Greek Fire was indeed a weapon to be feared.[10]

Sailors
Harnessing Wind for at Least 6,000 Years

The oldest image of a boat with a sail is on a ceramic jar found near the west bank of the Nile River in Egypt, dating back to the fourth millennium BC. A ceramic disc found in Kuwait is approximately 2,000 years older than that and shows a ship with a mast, though not a sail.[1] Sails were used with large canoes and with both reed and wooden boats.

In early sailing vessels, a single square sail suspended from a vertical pole (called a *mast*) by ropes (called *rigging*) moved the boat in the direction the wind was blowing. To travel against the wind, the crew took down the sail and deployed oars, putting rowers (whether willing crew or slaves) to work.

Sailors eventually learned how to sail at right angles to the wind by angling (or *trimming*) the sail to redirect air flow. From this, an ability to sail *against* the wind—zigzagging across the wind to reach an upwind destination—evolved in multiple sailing cultures. To accomplish this, a vessel needed enough *lateral plane* underwater from its hull, keel, or movable *foils* (rudder, centerboard, daggerboard, or leeboards) to resist *leeway*, or being blown sideways. Sailors began to experiment with multiple sails in various combinations of shapes and sizes on multiple masts to see which would move the ship forward the fastest in given wind conditions, and which would perform best in downwind or upwind sailing.

L.Verschuier/Rijksmuseum

The arrival of King Charles II of England in Rotterdam, May 24, 1660. The square-rigged ship at left fires a cannon salute. The gilded Royal yacht in the foreground deploys a leeboard from her side amidships to reduce leeway (side-slipping), as do the two smaller sailboats at right. A variety of sailing rigs (sprit, lug, and gaff rigs) are visible.

A ship carried a large inventory of sails so that the right combination would always be available. Fore-and-aft sails—which attached to a mast or stay along their leading edge—proved more *close-winded* (better able to sail upwind) than square sails, which set from horizontal *yards* that pivoted around a mast. Many a square-rigged ship was wrecked because of its inability to "claw to windward" off a lee shore in a storm.

The first sails were made of natural materials—papyrus mats in Egypt, bamboo mats in Asia, or thin leather in northern Europe.[2] By the Middle Ages, woven flax or hemp was the most common material,[3] giving way to cotton in the mid-eighteenth century. Sailmakers had to balance the strength and durability of a fabric against its weight. A good sail must resist the effects of sunlight and salt air. It should not stretch or change shape when wet. Today, most sails are made from synthetic fibers including polyester, Kevlar, and carbon fibers.[4]

For ships sailing long distances, speed was always primary. The quicker the voyage, the less the crew had to be paid or fed. Cargoes could lose value in transit, and a ship's captain did not get paid until the voyage was completed. The more voyages a ship could make in a year, the more money the captain and the shipping company could earn. The California Gold Rush and competition in the China tea trade increased demand for speedier sailing ships.[5] The clipper ship, developed in the mid-nineteenth century, was designed for speed under sail.[6]

The average clipper could sail from New York around South America to San Francisco in about 120 days, as compared with the six-month land journey over the Oregon Trail, and the fastest "extreme" clippers were able to make the trip in as little as 100 days. The record of 89 days, set by the *Flying Cloud* and the *Andrew Jackson* in 1854,[7] was unbroken until 1989, when *Thursday's Child*, a high-tech racing yacht, arrived in San Francisco 81 days after leaving New York.

This Antonio Jacobsen painting shows the American clipper ship *Flying Cloud* at sea under full sail. A full-rigged ship is a sailing vessel with three or more masts, all carrying square sails. More generally, however, the word ship refers (to this day) to any oceangoing vessel—showing that the language of the sea can be at once precise and baffling. Clipper ships reached their zenith in the mid-nineteenth century before being chased from the seas by steamships.

As steam-powered ships became safer and more reliable, they replaced sailing ships for travel and commerce. Lack of wind did not delay steamships, and they required fewer crew members, making them cheaper to operate. Today, most sailing vessels are used only for recreation, but rising fuel prices and limits on carbon emissions may change that. Rolls-Royce Holdings and B9 Shipping are working to develop a 330-foot (101-meter) modern-day cargo-carrying clipper ship with three 180-foot-tall (55-meter) sails and biomethane engines.[8] Automated systems can now perform much of the work needed to raise, lower, and reposition sails, and it takes only one person to operate the *Maltese Falcon*, a 290-foot (88-meter) luxury yacht built in 2006.[9]

Sails for Land Vehicles

An iceboat on a lake in Poland.

Mafiozo

Sails have been used to propel vessels other than ships. Ice boats, developed in the U.S. in the mid-1800s, were mounted on sled runners or skis and used to move people and goods across frozen surfaces. Land sailing, begun in China around AD 550, uses a sail to move a wagon or other wheeled platform across a flat surface. In the 1860s, some settlers of the American West used "windwagons" to travel across the Great Plains.

Sailing in Space

Artist's depiction of the Japanese IKAROS space probe in flight.

Andrzej Mirecki

Solar sails use solar (radiation) pressure from stars to drive a vehicle. The sail is a giant reflective sheet. When light particles from the sun bounce off the sheet, the energy created moves the vehicle forward much as wind powers traditional sailing ships.[10] In June 2010, Japan deployed the first successful solar sail spacecraft, called Ikaros (Interplanetary Kite-craft Accelerated by Radiation of the Sun). Ikaros also uses thin-film solar cells on the sail to generate electricity for additional power. The goal is for Ikaros to be the first successful space mission powered only by sunlight.[11] Meanwhile, NASA scientists are exploring development of a land-sailing rover for use on Venus. The rover would take advantage of the planet's low-level surface winds and relatively flat surface.[12]

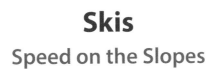

Skis
Speed on the Slopes

The oldest ski unearthed to date was found in snowy northern Russia, near Lake Sindor.[1] It dates back to approximately 6000 BC, and cave drawings in the area of its discovery suggest that humans used skis as early as the last Ice Age, approximately 10,000 years ago.[2] Some experts believe skis were used in China at approximately the same time.[3] Skiing is far from a modern idea.

Skis may have developed from snowshoes in areas where the snow was deep and compact;[4] ski poles may have begun as walking sticks that snowshoers used for balance.[5] Like snowshoes, skis were an excellent vehicle for crossing frozen wetlands and marshes and for winter hunting. Their use spread widely across Central Asia and Europe.

Early skis varied in length and were used with only one pole. Around AD 1600 in Norway, skiers wore a long smooth board coated with tar on one foot and a shorter, fur-bottomed board on the other. The skier would use the foot with the shorter board, which had better traction, to push forward, then glide on the other foot.[6]

Skis were used for military purposes in Scandinavia in the late seventeenth century, and King Karl XII of Sweden organized companies of ski soldiers for battle in 1716 and 1718. The Norwegian Army established permanent companies of ski soldiers in 1747, and skiers continued to be an important element of that country's military until

The Ski Museum

A depiction of Birkebeiner skiers carrying Prince Haakon of Norway to safety during the winter of 1206. The prince grew up to be King Haakon IV and ended Norway's long-running civil war.

Norway's union with Sweden in 1847. Skiing competitions with cash prizes were held to identify and recruit the best young skiers.[7] Today's Winter Olympics biathlon competition, which combines cross-country skiing and rifle shooting, reflects the skills these soldiers needed.

Modern skis incorporate a number of improvements over their ancient predecessors. In the mid-nineteenth century, Norwegian wood-carvers in the Telemark province invented the cambered ski, which arches upward in the center when unloaded so that the wearer's weight will be more evenly distributed along its length. Earlier skis had to be thick enough to glide across the snow without dipping in the middle, where the majority of the skier's weight was centered; cambering allowed for thinner, lighter skis that were easier to maneuver. In 1882, using modern carbon-steel tools, Norwegians began making skis from hickory, a hard, tough wood, which allowed for thinner, more flexible skis.

The early twentieth century brought segmented steel edges to help skis grip the snow better, three-layer laminate skis, and an aluminum ski. In 1946, the Gomme ski, produced in England, was the first to combine disparate materials: a laminated wood core sandwiched between a plastic upper layer and a metal bottom layer. The first successful plastic (fiber-reinforced plastic, or fiberglass) ski, the Tom Mailer, was introduced in 1959. By the 1970s, manufacturers were mixing Kevlar, carbon fiber, ceramic fiber, and other high-strength materials into fiberglass to build stronger, faster, more maneuverable skis.[8]

A skier carving a turn.

Skiing did not emerge as a leisure activity until the mid- to late-nineteenth century, when ski clubs and competitions became popular. The 1880s saw a shift from cross-country (Nordic) to downhill (alpine) skiing as people discovered the adrenaline rush of whizzing down a steep mountain slope.[9]

Today, both cross-country and downhill skiing are popular, and ski events such as the slalom draw thousands of spectators to the Winter Olympics. Ski resorts cater to people of any age and level of skill, from beginners to experts, and cross-country skiing remains an enjoyable way to travel through a winter landscape.

Austrian soldiers on skis during World War I.

Horses
To Ride Like the Wind

Until the invention of the steam engine, horses were the fastest, most reliable mode of travel on the planet.[1]

No one knows exactly how long people have been riding horses or using them to pull vehicles. Images of horses appear in cave paintings as early as 30,000 BC, but experts believe that the horses in the paintings were hunted for meat and skins, and that humans did not begin keeping or riding horses until much later.[2]

Researchers disagree on when horses were first domesticated and even on what "domestication" means. Is it simply taming wild animals and conditioning them to live with humans? Is it training an animal to obey commands and do useful work, like pulling a cart? Or is it actively controlling the breeding of new generations?

There are several lines of evidence for domestication of horses: changes in the skeletons and teeth of ancient horses; expansions of their geographic range; and the discovery of artifacts and images including equipment used with horses and pictures of horses being used for riding, pulling a chariot, dragging a plow, or other work.

Evidence shows that the Botai people in what is now Kazakhstan began domesticating horses approximately 5,500 years ago, and that they rode their horses and drank mares' milk in addition to eating horse meat.[3] People in Mesopotamia (what is now Iraq and parts of Syria, Iran, and Turkey) seem to have used horses as pack animals beginning

JWPetty51

Leonardo Da Vinci's sculpture
of a horse and rider.

about 3000 BC and were hitching horses to chariots by about 2000 BC.[4] In 1955, remains of two chariots and six horses dating to approximately 1100 to 700 BC were found in China's Western Zhou Chariot Burial Pit.[5]

Early research suggested that one group of people began raising and riding horses and the practice spread from there, but recent DNA analysis has shown that domestication of horses occurred in multiple locations and times across Europe and Asia. No one region or people can take full credit for this accomplishment.[6]

How horses were domesticated is also a matter of debate. Perhaps when people killed adult horses for food, they kept the young foals as pets, and the foals grew up feeling comfortable around humans.[7] Or perhaps wild horses were "broken" much as cowboys in the Old West tamed wild mustangs, gradually asserting mastery until the horse became complacent and obedient. Regardless of the method, horses soon became an indispensable means of travel, whether ridden or used to pull a chariot or a cart full of household or trade goods.

Horses were not the only domesticated animal used for these purposes. Mules, donkeys, oxen, reindeer, elephants, llamas, and camels have been used for peaceful and military travel or to farm the land. Farmers in many countries depend on Asian water buffalo to pull plows, wagons, and carts, and these animals are essential to rice farmers.[8] Dogs and goats

Przewalski horses, a subspecies once extinct in the wild, have been successfully reintroduced to their native habitat, the steppes of Mongolia.

Foto Halo/Thinkstock

can pull light carts, and sled dogs were, until recently, used extensively for travel in arctic regions. In short, humans throughout the world have domesticated animals common to their geographic regions and used them for assistance, selecting the ones best suited to the task at hand.

A Horse Extinction

Horses died out in the Western Hemisphere at the end of the last Ice Age, around 8000 BC, due to climate change and overhunting. Horses on the Eurasian continent may have avoided this fate because they were protected and fed in captivity. They were then reintroduced to the New World by Spanish explorers and conquistadors.

The Charge of the Elephants

A late twelfth or early thirteenth century relief from the Bayon temple in Angkor, Cambodia, depicting the Khmer army going to war against the Cham.

In its day, no weapon was more formidable than the war elephant.

People began taming wild elephants around 2000 BC, and their first use in battle dates to around 1100 BC. Their huge size terrified soldiers who had not seen an elephant before, causing them to break ranks and flee. A group of charging elephants could trample the enemy and easily smash through fortifications, and their thick hides made them difficult to injure or kill.[9] Soldiers mounted on the elephants' backs had a high platform from which to launch arrows or spears. On the other hand, when injured or frightened, elephants could run amok, harming soldiers in their own armies, and their large size made them highly visible targets.[10]

The best-known use of elephants in battle was by the Carthaginian general Hannibal in the Second Punic War (218–201 BC) against the Romans. Determined to surprise the enemy by attacking from the rear, Hannibal led his troops and 37 elephants across the Alps. What is less well-known is that most of the elephants perished in the mountains; only the few survivors rode into the battle.

The use of elephants in war did not end with the Roman Empire. In World Wars I and II, soldiers in India and Southeast Asia used elephants to haul heavy equipment, supplies, and artillery through thick jungles and across rivers— places where vehicles would get stuck—to build roads and bridges. Elephants served a similar purpose in the Vietnam War. Today, trucks, helicopters, and amphibious vehicles have replaced elephants in the military.[11]

The Wheel
Keep on Rolling

A wheel isn't a vehicle, but the invention and development of the wheel have been central to many forms of transportation through the ages. Without the wheel, vehicles from carts to bicycles, cars, trucks, and buses would have been impossible. Even airplanes need wheels when taxiing, taking off, and landing.

The oldest wooden wheel discovered to date is the Ljubljana (Slovenia) Marshes wheel, believed to be between 5,350 and 5,100 years old.[1] (Stone potters' wheels of similar vintage have been found in Mesopotamia, but there is no evidence of stone wheels being used on vehicles.) Archaeologists have found evidence of wheeled vehicles dating from approximately 4500 BC in Mesopotamia, the Indus Valley, the Northern Caucasus, and Central Europe. The earliest well-dated depiction of a wheeled vehicle (a wagon with four wheels and two axles)

The earliest vehicle wheels were solid wooden disks.

John O'Neill

is on the Bronocice pot, excavated in southern Poland and dated about 3500–3350 BC.

No matter where the wheel first originated, it spread quickly across Asia and Europe, most likely in the form of the chariot. By 1200 BC, wheeled vehicles were in use as far east as China and as far north as Scandinavia.

In its simplest form, the wheel was a solid wooden disk secured by wooden pins to a round axle, a wooden shaft that joined two wheels together. The cart or other vehicle platform sat atop one or more axles. The hole in the wheel allowed the wheel to revolve around the axle, while the axle itself did not rotate. Thus, while the wheels turned to move the vehicle forward, the vehicle itself stayed steady atop the axles. The first solid wood wheels were heavy and cumbersome. Eventually, sections were carved out of the disk to reduce weight, and spoked wheels came into use about 2000 BC. Spoked wheels were lighter, and vehicles using them could move faster.

National Museums Scotland

Remains of a wooden wheel found in the Drummond Moss swamp on the Blair-Drummond estate in Scotland. This is considered to be the first evidence of wheeled transport in Britain.

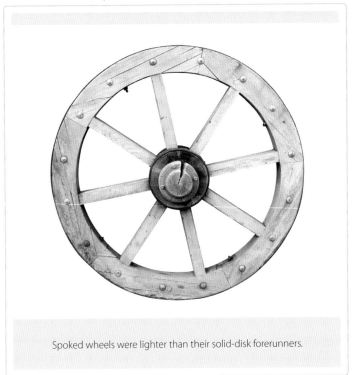

Churhai Cao/Thinkstock

Spoked wheels were lighter than their solid-disk forerunners.

Wheels and axles made it easier to move heavy loads on carts pulled by people. Carriages, specifically built to carry people, weren't commonly used until domesticated animals—such as horses—were available to pull them. In addition to needing a strong animal to pull it, a wheeled carriage also needed a smooth road on which to travel. Wheels don't work well in sandy, rocky, or uneven terrain or in deep snow or mud. Wheeled vehicles were much less common in parts of the world with unfavorable terrain.

European, Asian, and African cultures seem to have quickly found many uses for the wheel once introduced to it, but the same was not true in the Western Hemisphere. For example, the Olmec people of Mexico left behind what appear to be children's toys with wheel-like worked stones, but large-scale development and use of the wheel never occurred. Experts attribute this to the absence of large domesticated animals that could be used to pull wheeled vehicles. Horses became extinct in North and South America around 12,000 years ago, and didn't return until Spanish explorers reintroduced them in the late fifteenth and early sixteenth centuries. The Incas used llamas as pack animals and for food, but llamas are smaller than horses and not suited for pulling heavy loads.[2] A hand-pulled travois worked better on grasslands, sandy soil, or steep hills than a wheeled cart.

Evolution of the Wheel

Modern vehicles still run on pairs of wheels connected by an axle, but the resemblance of these wheels to their ancient predecessors ends there.

Around 1000 BC, the Celts began strengthening chariot wheels by surrounding them with iron rims. In 1802, G.F. Bauer received a patent for the first wire tension spoke, a wire that ran from the central hub, through the rim, and back to the hub, making wheels lighter and stronger. Solid rubber tires replaced wooden wheels, only to be replaced in turn by air-filled pneumatic tires beginning in 1845.[3]

Early wheels rotated directly around the axle, creating frictional heat, resistance, and wear. In 1869, Jules Suriray fitted ball bearings into the wheel hubs of a bicycle that then won the Paris–Rouen road race.[4] Adding ball bearings reduced friction around the axle, making the wheels turn more smoothly and allowing the rider to travel farther with each rotation of the pedals.

Newer, lighter materials, better design, and improvements in ball-bearing technology have continued to make wheels stronger and more efficient.

Chariots
Shock and Awe on the Battlefield

Anyone who has seen a movie about Roman gladiators or studied the history of Egypt has seen a chariot. These lightweight vehicles had two spoked wheels and were pulled by two or more horses. The first chariots were little more than an open platform with a waist-high guard in front to keep the driver separated from the horses. The charioteer stood while driving.

Chariots were a technological advance over four-wheeled carts or wagons pulled by donkeys or oxen. Horses were much faster, and wheels with spokes—rather than the solid wooden wheels used on early carts—made the chariots much lighter and easier to maneuver.

We associate chariots with the early Egyptians, Greeks, and Romans, but they may have been developed on the Eurasian steppes, along the border between Russia and Kazakhstan, more than 4,000 years ago.[1] By 1500 BC, chariots had spread into Egypt, Greece, and parts of Asia and northern Europe.[2]

The chariot has been described as an "early primitive tank,"[3] and its primary use was in warfare. In 1286 BC, Rameses II of Egypt sent 5,000 charioteers to overcome the Hittites. Greeks used chariots to conquer Troy, and there are accounts of chariots being used in China and India.[4] By providing a mobile platform, the chariot changed the way war was waged. Chariots enabled soldiers to travel quickly to the most

William McKelvie/Thinkstock

A stone relief of Ramses II fighting from his chariot at the battle of Kadesh in 1274 BC.

crucial areas of a battle, providing reinforcements where they were most needed. They served as platforms for archers, putting them above the heads of foot soldiers. Although it was difficult to shoot an arrow accurately from a moving chariot, a large group of chariots could overwhelm an enemy through the sheer quantity of arrows they fired. Imagine the shock of foot soldiers when first faced with horse-drawn chariots racing through their ranks, shooting arrows as they went! Charioteers were also armed with spears or short swords that could be used to stab fleeing enemies in the back.[5]

Charioteers enjoyed a much higher status than foot soldiers, but the job was expensive to hold. The royal stables provided the horses and five attendants that the charioteer had to equip at his own expense. He also had to buy and maintain the chariot at a cost that only the wealthy could afford.[6]

By approximately 1000 BC, horseback riding had become more common, and military chariots were being replaced by mounted cavalry. But chariots continued to be used for hunting; for ceremonial transport of pharaohs, high priests, and other dignitaries; and for sport racing. One historian notes that chariots were ideally suited for hunting lions.[7]

The first chariot race in the Greek Olympic Games was about 680 BC, and the Romans held chariot races in the Circus Maximus for crowds of up to 150,000 people.[8] In April 2009, as part of its celebration of 2,762 years of existence, the city of Rome staged a chariot race at the old Circus Maximus, using riders on bicycles instead of horses to pull the chariots.

In this image of a pharaoh driving a chariot, an attacking cheetah is at his side, and ahead of him runs an infantryman with longbow and arrows.

New York Public Library

Chariots as High Tech

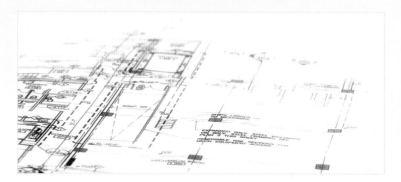

Today we see chariots only in movies, on TV, or in historical reenactments, but what we think of as a vehicle of ancient history was a technological marvel in its time.

A *Nova* documentary, first aired on PBS in February 2013, follows a team of archaeologists, engineers, woodworkers, and horse trainers as they build and test two accurate replicas of Egyptian royal chariots. Their research indicates that chariots incorporated a large number of advanced features, including (in addition to spoked wheels), springs, shock absorbers, anti-roll bars, and a rearview mirror. Bella Sandor, an engineer from the University of Wisconsin, described the chariot as "the world's first high-performance vehicle" and likened the level of design to the engineering standards of 1930s-era Buicks.[9]

Aerial Lifts
In Use More than 2,000 Years

Chairlifts at ski resorts are the best-known aerial lifts in use today, but aerial ropeways were developed centuries before skiing became popular.

The first known use of aerial ropeways occurred in the mountains of China, Japan, and northern India around 250 BC. Travelers could cross a deep ravine or river gorge by suspending themselves in a harness dangling from an overhead rope and pulling themselves hand over hand along the rope. Eventually the harness became a basket.[1]

Texts from the fourteenth and fifteenth centuries show ropeways in use in Japan and Europe. One describes how a Japanese emperor used a ropeway over a valley to escape his enemies. A European weapons catalog from 1405 refers to a ropeway, and a book published in 1411 contains a drawing of one. In 1615, Fausto Veranzio, of Venice, designed a passenger ropeway with a wooden box suspended from pulleys. Passengers sat in the box and pulled themselves along the overhead cable. In 1644, a Dutchman named Wybe Adam built a continuously circulating ropeway to carry materials up a hillside to a fortress construction site. Monasteries built atop high peaks used ropeways to transport supplies and visitors.[2]

All such early ropeways were powered by humans, animals, water, or gravity, and their lengths and carrying capacities were limited by the strength of the available rope. Development of wire cable, steam engines, and electric motors greatly increased the potential length and altitude gain of an aerial ropeway.

People discovered the thrill of skiing downhill by the 1880s,[3] but that brief thrill had to be preceded by a long, hard climb to the top of a mountain. A few resorts in Europe solved the problem by building small railways, but it wasn't until 1908 that a water-powered rope tow in Schöenaich, Germany, became the first lift built for winter sports. More ski areas were building rope tows by the 1920s, often from used

Paula Grey

Riding a tramway to Sandia Peak outside Albuquerque, New Mexico.

Record Breakers

The Cableway at Merida, Venezuela.

The world's highest-altitude aerial tramway is the Cableway of Merida, Venezuela. This system of four aerial trams connects the city of Merida with Pico Espejo (Mirror Peak) in the Sierra Nevada de Mérida mountains. The tramway ascends from an altitude of 5,380 feet (1,640 meters) to 15,633 feet (4,765 meters).

The tallest support pillar on a tramway is at Kitzsteinhorn in the Austrian Alps. Built in 1966, it is 372 feet (113 meters) tall.

Sandia Peak, in New Mexico, offers the longest-duration passenger aerial tramway ride in the world. The tramway takes 15 minutes to ascend from ground level to the peak, traveling a distance of 14,655 feet (4,467 meters).[7]

automobile parts.[4] These early tows, which pulled skiers uphill standing on their skis or sitting on a toboggan, were uncomfortable and occasionally dangerous.

When William Averell Harriman, chairman of the Union Pacific Railroad, developed America's first ski resort in Sun Valley, Idaho, in the 1930s, he wanted a safer, more comfortable way to lift skiers 2,000 feet up the slopes. James Curran, a young bridge designer working for the railroad, had previously designed a wire-based system for moving bales of bananas from loading docks to boats, and he thought the same approach might work for people. He replaced the hooks that held the bananas with chairs, and in 1936, his first working chairlift was installed at Sun Valley. By the 1960s, chairlifts had become standard equipment at ski resorts worldwide.[5]

Gondolas and aerial tramways are large aerial lifts covering long distances. In a gondola, small passenger cars, or carriers, are suspended at regular intervals from a cable loop and move continuously from the bottom of the mountain to the top and back down again. When a fully loaded carrier reaches the top of the mountain, workers slow it down and detach it. Passengers get out, new passengers climb in, and workers reattach the carrier to the downhill leg of the cable. The detaching, unloading, reloading, and reattaching is repeated at the bottom of the lift.

An aerial tramway uses large passenger cabins and travels high above the ground. Unlike gondolas, aerial tramway cars are not detachable, and only one car can travel on a cable at a time. The carrier travels up the mountain and then reverses direction to descend on the same cable leg.[6]

Camels
Ships of the Desert

Large swaths of Africa, Asia, and the Arabian Peninsula are desert. The world's three largest subtropical deserts—the Sahara, Kalahari, and Arabian—cover parts of what are today Egypt, Morocco, Tunisia, Iraq, Jordan, Kuwait, Saudi Arabia, Botswana, South Africa, and Namibia. Desert terrains include gravel plains, sand, dunes, and rocky highlands,[1] and a foot traveler in high heat wouldn't get far across any of these. Horses sink into sand, and wheeled carts are useless. But nature provided one animal perfectly designed for desert travel: the camel.

There are two surviving species of camel, and humans have domesticated both. The dromedary is a one-humped camel found in parts of North Africa and the Middle East. Bactrian camels, which have two humps, are found in central Asia, northern China, and Mongolia.[2]

The hump is a boneless mass of fleshy tissue that stores fat; it stiffens when a camel has plenty to eat and drink and shrinks as the camel depletes its resources.[3] Camels have astounding stamina and can go several days without food or water. A camel can walk at about 12 miles (19 km) an hour for up to 18 hours without resting, and can carry up to 600 pounds (272 kilograms).[4]

Camels have adapted to temperatures that reach 140° F (60° C) in summer. Their thick fur protects them from sunburn. A double row of eyelashes protects their eyes from the sun, and a third eyelid that closes sideways keeps out dust and sand. Brushy hairs in a camel's nose and

Matejh photography/Thinkstock

A caravan on rocky terrain.

the ability to pinch its nostrils shut also keep out sand.[5] Tough, leathery skin protects the bottom of a camel's feet, and each foot has two toes with a flap of loose skin between them. When the camel walks on sand, the skin spreads out to create a kind of snowshoe that keeps the animal from sinking.[6]

Camels are herbivores (plant eaters) and get most of the moisture they need from their food. Their tongues and mouth membranes are tough enough to keep them from being injured by the thorns and prickly spines of desert plants that other animals can't eat. They can also find water by smell.[7]

By about 200 BC, camels were being used as pack animals on trade routes carrying goods from China, India, and the Middle East to Europe and back again.[8] Centuries later, devout Muslims making the pilgrimage to Mecca from Cairo or Damascus traveled by camel.[9]

Just as westward-bound travelers on America's Oregon Trail created wagon trains in the nineteenth century, desert herdsmen and traders formed camel trains composed of hundreds of animals carrying goods and passengers. The camel puller, responsible for a group of animals, tied a rope from the nose ring on each camel to the saddle of the camel in front of it, creating a file of up to 40 animals. Several files,

The Silk Road

The best-known use of camel trains was along the Silk Road, a network of trade routes stretching west from China approximately 4,000 miles (6,437 km) through Asia and northern Africa to the Mediterranean. The original

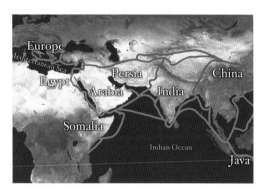

The Silk Road, with land routes in red.

route, developed in the second century BC, connected China with Constantinople, the capital of the Roman Empire. Silk was a highly prized commodity, and only the Chinese knew the secret of making it, but many other goods were carried along the Silk Road. Over time, the main route incorporated other trails and developed new branches connecting it to major cities and trading centers.

Goods carried along the Silk Road included African gold and ivory, Indian pepper, Persian metal crafts, Baltic amber, and Scandinavian furs.[14] Ceramics, spices, gems, incense, fruits, and flowers were all traded.[15] Trading posts and settlements grew up along the route.

The Silk Road carried more than trade goods. Inventions such as gunpowder, the printing press, the spinning wheel, and the mechanical clock were passed from one civilization to another, as were political and religious movements such as Buddhism and Islam. And like today, infectious diseases accompanied travelers and found new populations to afflict.

each led by a puller, traveled side by side or in a single long line.[10] Each caravan was led by an experienced guide, skilled at reading the desert.[11] The last camel in each file wore bells on a stick attached to its saddle, and if the camel puller stopped hearing the bells from his file, he knew that one or more camels had separated from the group.[12]

Today, trucks carry goods along most desert trade routes, but camels are still used in remote and hard-to-reach areas. In 2006, camel trains were regularly carrying salt from the Taoudenni salt mines in northern Mali across 450 miles (740 km) of desert to the city of Timbuktu.[13]

Camels in America and Australia

Australian Army Camel Corpsmen wrestling with a recalcitrant beast.

In 1856, the United States Army established the Camel Corps and imported 74 camels and their handlers from Turkey for use in the dry desert of the American Southwest. Army planners thought the camels would help in battles with Native American and Mexican fighters. From 1859 to 1860, the camels were used by surveyors mapping the Big Bend area of Texas, and during the Civil War, camels carried supplies to southwestern Army posts. But the use of camels never really caught on. The animals could be stubborn and aggressive and would sometimes bite, spit, and frighten the horses. In 1865, the Camel Corps was disbanded. Twenty-eight camels were sold to the city of Los Angeles for carrying mail and baggage, and the rest were sold to miners and ranchers. Some later ended up in zoos, circuses, and traveling menageries, while others escaped into the desert to live wild. Topsy, the last documented offspring of the Army camels, was captured in California and taken to live in a zoo, where she died of old age in 1935.[16]

Around the time the U.S. Army was establishing its Camel Corps, the Australian government began importing camels and drivers to help explore its inland deserts. By 1900, there were more than 6,000 camels in Australia carrying food, supplies, and equipment for engineers who were surveying and building the Overland Telegraph Line and railways. Camels also carried supplies and mail to remote settlements. By the 1930s, however, railways and roads had eliminated the need for camels. Many were turned loose, and as of 2013, Australia was still home to an estimated 300,000 wild camels.[17]

Dogsleds
Driving Through the Snow

Long before European explorers and settlers reintroduced horses to North America, the indigenous peoples of its northern regions were hitching dogs to sleds. And they were not alone.

Arabian literature from the tenth century describes the use of sled dogs in Siberia, as do the thirteenth-century writings of Marco Polo. Archaeologists have found evidence of dogsleds in Canada dating back four to ten centuries. *Historic Navigations*, a book published in 1675, contains an illustration of a dog in harness pulling a sled.[1]

Dogs made good sled animals. They were light enough to run on top of packed snow that horses would sink through, and dogs that were native to northern regions were adapted to the frigid winter temperatures and heavy snow. By the mid-eighteenth century, the use of dogsleds to carry mail and freight had become common among colonists in Canada. French-Canadian soldiers used dogsled teams during the French and Indian War, and by 1775, dogsleds were essential to the Northwestern fur trade.[2]

In addition to roads, people living in the far north forged dogsled tracks. These trails of hard-packed snow connected settlements to one another and to major centers of trade, enabling travel and commerce to continue through the long winter months.

Dogsleds were made of thin planks of oak or birch lashed together with deerskin. The front ends turned up to protect the contents from flying snow kicked up by the dogs. Larger sleds, used for freight, could be up to 12 feet (3.7 meters) long. Various accounts estimate the weight that a team of two dogs could haul as anywhere from 200 to 500 pounds (91 to 227 kg), depending on the dogs and the distance to be traveled.[3] Smaller sleds with sides made of moose skins stretched over planks were designed to carry one or two fur-wrapped passengers.

A dogsled on the Denali trail.

No matter the length, dogsleds were narrow, typically about 16 inches (0.4 meter) wide. When there was no trail for the dogs to follow, the driver walked ahead on snowshoes to pack the snow so the dogs could walk on a solid surface, and the sled had to fit on that narrow track. When a trail was already well established, the driver walked or ran behind the sled, occasionally standing on the back to ride for a few minutes. Teams could travel at an average speed of about 4 miles (6.4 km) per hour, the speed of a brisk walk.[4]

The use of airplanes to carry mail and supplies, beginning in the late 1920s, led to a decline in the use of dogsleds for long-distance hauling,

Lomen Bros, Nome/Library of Congress

The dogsleds for the Stefansson-Anderson Canadian Arctic
expedition were built in Nome, Alaska, in 1913.

but they were still widely used for local travel and day-to-day work. During World War II, the Alaska Territorial Guard, known as the Eskimo Scouts, used dogsleds to patrol western Alaska and safeguard the coast from a possible Japanese invasion. It took the widespread introduction of the snowmobile to significantly diminish dogsled use.[5]

Dogsleds are still in use today, and many communities host winter dogsled races to keep the tradition alive. Some people believe that nothing will ever totally replace the dogsled. Joe Redington, known as the father of the Iditarod, said, "I've seen snow machines break down and fellows freeze to death out there in the wilderness. But dogs will always keep you warm, and they'll always get you there."[6]

The Iditarod

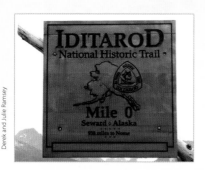

A marker for the start of the Iditarod Trail.

In 1925, Dr. Curtis Welch, the only doctor in Nome, Alaska, faced a serious threat. Diphtheria, a lethal respiratory illness, was spreading among the town's children. Fearing an epidemic, Dr. Welch ordered a quarantine. But what he really needed was an antitoxin serum, and the closest source was Anchorage, more than 1,000 miles (1,609 km) away. Nome's harbor was frozen over, and the weather made air travel impossible. The governor of Alaska recruited volunteer drivers and dog teams to deliver the desperately needed serum to Nome.

The serum was sent to the town of Nenana by train and then run the nearly 700 remaining miles (1,127 km) by a nonstop relay of 20 drivers and 150 dogs. The serum arrived in five and a half days, just over half the time the trip by dogsled normally took.[7]

In 1967, volunteers commemorated the hundredth anniversary of Alaska becoming a U.S. territory by clearing a portion of the historic freight route called the Iditarod Trail and staging a dogsled race. Their inspiration was the 1925 race against time to deliver the diphtheria antitoxin. In 1973, the entire trail was cleared for the first long-distance Iditarod Race. Today, the race, held every March, is Alaska's best-known sporting event. Dogsled teams from all over the world race between checkpoints from Anchorage to Nome, covering a total of almost 1,150 miles (1,851 km).

Chinese Treasure Ships
Supertankers of the Ancient World

Marco Polo was a Venetian merchant who traveled to China in 1271 and lived at the court of the great Chinese ruler, Kublai Khan, for 17 years. While there, he traveled to parts of Asia never before seen by Europeans. After returning to Venice, he published a book he called *The Description of the World.* The book was printed in French, Italian, and Latin and read throughout Europe, but most readers refused to accept that it was true. They soon gave it another title, *Il Milione,* or *The Million Lies.* When Marco Polo was dying in 1324, friends begged him to admit that his book was fiction, but he refused, saying, "I have not told half of what I saw."[1] One of the wonders Polo's readers refused to believe was his description of enormous Chinese sailing ships, called treasure ships because they brought back treasures from other lands.

The Chinese had long sailed the rivers of their territories. They became expert at digging canals for irrigation, flood control, and transportation. By 600 BC, they had developed a sea trade along the coast,[2] and by 87 BC, they had expanded their trade routes to include parts of India and Southeast Asia.[3] Written accounts from around AD 1225 describe 46 far-flung places where the Chinese traded, including Japan, Southeast Asia, the Indian Ocean, and Mediterranean ports such as Alexandria in Egypt, the Italian island of Sicily, and Andalusia in Spain.[4]

When Marco Polo reached China, the Chinese were far ahead of Europeans in shipbuilding and navigation skills. Their ships were far larger and carried much more cargo. The Chinese had begun using a centerline rudder for steering nearly a thousand years before the Europeans.[5] They began using a magnetic compass as early as 1117 and were the first to divide a ship's hull into compartments, which strengthened the hull and ensured that flooding could be contained in one area rather than sinking the entire ship.

By the early 1400s, the Chinese treasure ships had grown even larger than those Marco Polo saw. Between 1405 and 1433, Chinese admiral Zheng He commanded seven expeditions, each with a large armada of up to 317 ships. The largest ships were an estimated 440 to 538 feet (134 to 164 meters) long and 210 feet (64 meters) wide—four to five times the size of Christopher Columbus's flagship, the *Santa María*. Zheng He's mission was to establish ties with foreign rulers, impress them with China's wealth and power, and initiate trade.[6]

The display in the China Court of the Ibn Battuta Mall in Dubai suggests (but may exaggerate) the relative sizes of Admiral Zheng He's ships of the early fifteenth century and those of Columbus's later expedition across the Atlantic.

Lars Plougmann

Each of the 62 largest ships in the first armada had four decks and nine masts. The rest of the armada consisted of approximately 185 smaller ships that carried food and water, horses, soldiers, and goods to trade. It is estimated that the crew totaled more than 27,000.[7]

Archaeological evidence of these giant ships is rare, but in 1962, the 36-foot-long (10.9 meters) rudderpost of a treasure ship was found in the ruins of a boatyard in Nanjing. Using the proportions of a traditional

Chinese junk, archaeologists estimated that the rudderpost was built for a ship with a hull approximately 500 feet (152 meters) long. Another wreck, discovered in Quanzhou in 1973 and dated to the 1270s, has a pine keel over 100 feet (30.5 meters) long. Its 13 hull compartments contained remnants of spices, shells, and fragrant woods from east Africa.[8] These discoveries confirmed Marco Polo's descriptions.

A full-size model of a "middle-sized treasure boat" (63.25 meters or 207 feet long) of the Zheng He fleet at the Treasure Boat Shipyard site in Nanjing, built in 2005 from concrete faced with wooden planks.

Vladimir Menkov

How Far Did the Chinese Travel?

In 2002, British author Gavin Menzies published *1421: The Year China Discovered America*, asserting that Zheng's armadas and later Chinese fleets explored as far as North and South America, Siberia, Australia, and Antarctica. For proof, he cited ancient Chinese maps and texts that he claimed described these parts of the world well before European explorers discovered them.[9] For further proof, Menzies pointed to artifacts he claimed were wrecks of Chinese ships, including one found buried under a sandbank in the Sacramento River in California.[10] Menzies's book caused great excitement when it was published, but reputable experts have since disproved many of its claims.[11] There *is* proof, however, that Zheng He's fleet sailed as far as the Cape of Good Hope, the southern end of Cape Peninsula, Western Cape province, South Africa.[12]

Erasing History

Admiral Zheng He's voyages were an exceptional achievement, but there is little documentation of them in official Chinese history. Much of what we know today comes from *The Ming Tong Jian*, an unofficial history of the period; the *Li-Tai Thung Chien Chi Lan*, an imperial history compiled in 1767,[13] more than three hundred years after the voyages; and from records, monuments, and oral histories in the countries Zheng visited.

Why are Zheng He's voyages so little known in his own country? The answer seems to be politics. Emperor Zhu Di, who commissioned the voyages, died in 1424 before they were completed. His successors believed that the expeditions were too costly, harmful to the empire, and against the principles of Confucians, a powerful faction in the emperor's court. They believed that China had nothing to gain from relations with "barbarian" nations outside the Chinese empire.[14] Instead of continuing to explore, the new emperor imposed a ban on all maritime trade that lasted more than one hundred years.[15] Ports were closed, Zheng He's ships were burned, and the records of his voyages were destroyed.

Wheelchairs
Mobility for All

It seems only logical that once wheeled vehicles were developed they would be used to help move people who could not walk on their own. But no one knows who first got the idea of putting a bed or a chair on wheels to help the disabled.

An image on a stone slate from sixth-century China is believed to be the first depiction of a wheelchair. The earliest wheelchair designed expressly for that purpose is believed to be an "invalid's chair" made in 1595 for King Phillip II of Spain.[1] In 1655, Stephen Farfler, a 22-year-old paraplegic watchmaker, built his own self-propelling chair with two large wheels in the back and a single, smaller front wheel. The rider turned a hand crank connected to the front wheel to move the chair forward or backward. Farfler's invention looked more like a soapbox-derby car than a modern wheelchair.[2]

In 1783, John Dawson invented a more modern-looking wheelchair that became known as the Bath chair after the town of Bath, England, where many people went to bathe in the hot springs and mud baths that were supposed to restore health. The Bath chair had a high back, two large back wheels, and a smaller front wheel. Medical instrument catalogs began to advertise wheelchairs as a way to transport patients.[3]

The wooden Bath chair was heavy and uncomfortable, but improvements eventually led to lighter, more comfortable models. In 1869, a patent application showed the first wheelchair with rear push wheels and small front wheels, similar to those in use today. By 1881, inventors had replaced heavy wooden or metal wheels with hollow rubber wheels on metal rims and had added push rims that enabled the rider to move the chair more easily. Wheels with spokes were added in 1900.[4]

The first folding tubular steel wheelchair came about in 1932 when a friend of engineer Henry Jennings wanted a wheelchair that could fit easily into a car. This invention is still considered one of the most important twentieth-century advances in wheelchair technology, because it allowed disabled people to use their wheelchairs outside the home or hospital. Depending on a wheelchair no longer meant being confined!

Drawing of Stephen Farfler's self-propelled wheelchair, ca. 1655.

Canadian George Klein and a team of engineers invented the first electric-powered wheelchair as part of a project by the National Research Council of Canada to assist injured World War II veterans.[5]

Wheelchair racers at the April 2014 London Marathon.

Julian Mason

Early electric wheelchairs were standard wheelchairs with a motor attached. Later models incorporated the motor and batteries underneath the chair's seat and provided a smoother ride and better control by the operator.[6]

Today's wheelchairs are lighter, more comfortable and maneuverable, and easier to operate than earlier models. Owners of motorized wheelchairs can use a joystick, a head stick, or a sip-and-puff device to move and steer. Specialized wheelchairs can climb stairs or raise the rider to a standing position. And an outdoor power chair that moves easily through mud, sand, snow, water, and uneven terrain helps people enjoy outdoor activities.[7]

Many innovations in wheelchair design have come from their use in sports, in particular the international Paralympic Games. The

Paralympics started in 1948 with a competition in England for World War II veterans with spinal cord injuries, and has grown to an international event. Today, the Paralympic Games are held every four years in conjunction with the Summer and Winter Olympic Games. Athletes with disabilities from around the world compete in sports ranging from archery and fencing to tennis, hockey, and wrestling.

Wheelchairs for Everyone

Whirlwind International

Whirlwind International's 2010 RoughRider wheelchair.

An organization called Whirlwind International works to provide all-terrain wheelchairs to users in developing countries. These durable wheelchairs are designed to roll over curbs and rough ground without capsizing, fold easily, and can be repaired with commonly available tools and parts. Whirlwind International's goal is to give a wheelchair to every person who needs one, affording them as much personal independence as possible. Today their RoughRider all-terrain chair is used by 25,000 people over conditions that range from "muddy village paths to pot-holed urban streets."

Horse-Drawn Carriages
Traveling in Style

The Brown Brothers

One of the last horse-drawn carriages in New York City, 1917, photographed beside an electric trolley powered by a conduit between the rails.

After the fall of the Roman Empire, the roads the Romans had built throughout Europe deteriorated. As a result, few people traveled in wagons or carts. Walking or riding a horse was preferable to jolting about on potholed dirt tracks or getting stuck in mud or snow.

Roads improved by the early 1600s, however, and carriages once again came into use. They were available for hire in London as early as 1605, and by the 1650s, Londoners were complaining about traffic jams.[1]

Carriages also became more comfortable. A suspension built of leather straps (and later, metal springs) made the ride smoother, and enclosed carriage bodies protected passengers from the weather. Window openings were covered with glass after 1680, providing even more protection.[2]

Owning a private carriage became a status symbol. A wealthy aristocrat's most prized vehicle was a state or town coach emblazoned with the family coat of arms, and he might have as many as six or seven other carriages of various sizes for his family's varied needs.[3] The middle class had to make do with a one-horse two-wheeled gig, a pony and trap, or a donkey cart.[4]

Carriages ranged from large vehicles that could seat half a dozen people to small, one-person gigs. A large four-wheeled wagonette provided room to store food for a picnic or rods and guns for hunting. A dog cart included a ventilated space for transporting hunting dogs.[5]

Owning carriages was expensive, requiring at least one or two well-trained horses, a groom to care for them, a coachman to drive, a coach house, stables, and livery (a uniform) for the coachman. Aristocrats spent up to one-fifth of their annual income on these expenses,[6] and that did not include the initial cost of the carriage and horses.

Horses were selected for appearance more than speed or stamina. A well-trained horse knew how to "stand gracefully; to stop stylishly; to carry its head well...and to acquire the mandatory high-stepping action."[7] When a carriage required two or more horses, it was important that they match in size and color, look good together, and walk in tandem.

For the wealthy, owning a carriage was as much about image as transportation. Ladies in London for the social season (the period between Easter and July) spent afternoons riding through Hyde Park to show off their carriages, horses, and fashionable dresses.[8] Central Park in New York City was home to a similar afternoon promenade.

People who didn't own a carriage could rent one in London, as could visiting aristocrats. Livery stables would rent a carriage, horses, and coachman in livery by the day, week, month, or year at reasonable rates.[9] Driving a carriage through crowded city streets required considerable skill. By the 1900s, road traffic included "electric cars and trams, horse-drawn omnibuses, steam-driven lorries, petrol-driven motor-cars, bicycles, and all the old horse-drawn vehicles from barouches to baker's vans."[10] In thick fog, the driver could find it impossible to see the horses or the road. There were no headlights, only the candles in a pair of carriage lamps. People living outside the city often scheduled parties around the full moon to ensure safe driving for guests.[11]

Fancy horse-drawn carriages are rarely seen today. The Amish in parts of the U.S. still use horses and plain black buggies, and carriage rides are offered as tourist attractions in some cities. In 2014 and 2015,

A vintage carriage and couple in period dress.

Markus Schieder/Thinkstock

New York was debating whether to ban the popular carriage rides in Central Park. Supporters of the ban said the horses were overworked and should not be required to walk in city traffic, while opponents argued that the carriage rides were a tradition and a tourist attraction.

A March 2014 poll of New York voters found that 64 percent wanted to keep the carriages,[12] and the romantic rides through Central Park were still available as of 2016.

The Royal Carriages

The Queen of England's Gold State Coach in the Royal Mews.

David Crochet

The largest and most ornate carriage collection belongs to the Queen of England. In the late 1800s, the Royal Mews, Queen Victoria's stables, housed 32 coaches, more than 100 horses, and more than 125 staff ranging from the senior Master of the Horse down to footmen and stable hands. Each of the 13 coachmen had four different liveries to select from according to the formality of the occasion. "The most gorgeous and most costly was the full state livery worn on the most formal occasions, which consisted of a scarlet and gold striped frock coat, scarlet knee breeches, pink silk stockings, gold buckled shoes, a wig and a three-cornered hat decorated with pink ostrich feathers...."[13]

The number of royal carriages is still growing. In June 2014, Queen Elizabeth rode to open Parliament in a brand-new Diamond Jubilee State Coach, a gift from the Australian government for her eightieth birthday. It will join two other famous coaches: the Gold State Coach, which dates back to 1762, and the Scottish State Coach, built in 1830.[14]

Yachts
Boating for Pleasure

Boating for pleasure dates back to the ancient Egyptians. Pharaohs floated on the Nile River on royal barges while living, and after death, they were entombed with elegant barges to carry them through the afterlife.

The word *yacht* comes from the Dutch word *jagen*, meaning to hunt or chase. Originally, the word referred to fast, fully rigged sailing vessels sometimes used to chase pirates. In 1606, a 60-ton Dutch yacht, the *Duyfken*, was the first Western ship to land in Australia. Thirty-six years later a similar vessel made the first European sighting of New Zealand.[1] By the 1620s, hundreds of yachts were cruising Dutch waterways.

In 1642, with the British Civil War raging, the British royal family sent the future Charles II to the safe haven of an island off the coast of France. Sixteen-year-old Charles began sailing to pass the time, and later he went to Holland, where he continued to sail. When he returned to England to claim the throne in 1660, the Dutch East India Company gave him a 66-foot (20-meter) yacht, and he later commissioned several more. Meanwhile, Peter the Great, tsar of Russia, became interested in yachting after discovering the wreck of a yacht that had been owned by Ivan the Terrible a hundred years earlier. Interest in sailing for pleasure soon spread across Europe.[2]

In the late 1800s, yachts became a status symbol for wealthy Americans. Europeans continued to call sailboats used for recreation "yachts"

Martin Varsavsky

The luxury motor yacht *Ocean Pearl* at Tavolara Island off Sardinia, Italy.

regardless of their size, but Americans reserved the term for large, luxurious sail- or motor-powered vessels.[3] By the 1920s, financiers were commuting to Wall Street from their estates on Long Island and the Connecticut shore in high-speed launches, and rumrunners with aircraft engines (which often doubled as yachts) routinely outran Coast Guard revenue cutters throughout the Prohibition era to land illicit booze on secluded shores for speakeasies and wealthy tables.

Many wealthy Americans lost their fortunes in the stock market crash of 1929 and the Great Depression that followed, and interest in yachting diminished. That interest recovered in the postwar boom of the 1950s, however, and the introduction of fiberglass boats in the 1960s extended the concept of boating for pleasure to the burgeoning middle classes of America and Europe. For the first time, multiple copies of a boat could be built from a single mold; fiberglass boats were cheaper, more durable, and easier to maintain than wood. Yachting became boating.

In the financial boom of the 1990s, the emphasis shifted once again toward yachts for the wealthy and away from boats for the middle classes, more and more of whom were becoming two-income families with little leisure time. The competition among the rich to see who could boast the biggest yacht led to what are called *megayachts* or *super-yachts*—boats over 200 feet (61 meters) long.[4]

Megayachts can cost up to several hundred million dollars to build. Most have four or five decks above the waterline and one or two decks below. Luxurious staterooms and dining rooms, indoor Jacuzzis, swimming pools, helicopter landing pads, wine cellars, and movie theaters are among the amenities these yachts can provide.[5]

Most yacht owners can expect to spend 10 percent of the purchase price for annual maintenance. For multimillion-dollar yachts, this means yearly maintenance costs of hundreds of thousands of dollars. Large luxury yachts require a full-time staff, including a captain, engineers, stewards, chefs, and deckhands. A 220-foot (70-meter) yacht can consume 130 gallons (500 liters) of diesel fuel per hour just to run its shipboard systems while sitting still; at its cruising speed of 15 to

freeartist/Thinkstock

This steel sailboat on a Dutch waterway is a seaworthy yacht for a budget-minded sailor.

22 knots, it may burn more than 400 gallons (1,600 liters) of fuel per hour, costing thousands of dollars for a few hours of travel. Many yacht owners charter (rent out) their boats when they are not using them to help cover some of these expenses.[6]

The America's Cup

Almost as soon as people began sailing for fun, they began organizing races. The most famous yacht race is the America's Cup.

In 1851, Britain's Royal Yacht Squadron challenged the New York Yacht Club to take part in the Solent Races, and the New York club accepted the chal-

America's Cup catamarans dueling in the Louis Vuitton America's Cup World Series 2015 in Bermuda.

Yasuhiro Chatani

lenge. A year earlier, the club had commissioned a fast sailboat, the schooner *America*, which they sailed "on her own bottom" (as opposed to transporting her in a freighter) to England for the race.

The Royal Yacht Squadron decided that the *America* was eligible for only one of the Solent Races, the 53-nautical-mile (98 km) "All Nations Race" around the Isle of Wight, the prize for which was the Hundred Guinea Cup. The *America* crossed the starting line behind all 15 British yachts but quickly caught up and at one point forged a lead of several miles over her nearest competitor; when Queen Victoria, who was watching, asked who was second, the answer was, "Your Majesty, there is no second." The wind died as the *America* entered the Solent for the home stretch, allowing the yachts behind her to close the gap, but she still crossed the finish line a full eight minutes ahead of her nearest competitor. The New York Yacht Club accepted the prize and renamed it the America's Cup. Today the America's Cup Race is an international competition, and the cup itself, now called the "Auld Mug," is considered one of the world's most prestigious sporting trophies.[7]

Taxicabs
Wheels for Hire

Horse-drawn carriages for hire, called *hackney cabs*, were available in London as early as 1605,[1] and Nicolas Sauvage began offering carriages and drivers for hire in Paris in 1640.[2] In 1654, the British Parliament passed "An Ordinance for the Regulation of Hackney-Coachmen in London and the Places Adjacent," limiting the number of cabs, drivers, and horses in the city. For the first time, hackney drivers had to be licensed, and the London Court of Aldermen was charged with overseeing the cab business.[3]

The hansom cab, designed and patented in 1834 by Joseph Hansom, eventually replaced the hackney. The hansom was a light two-wheeled carriage with a low center of gravity, making it less likely to tip over on tight turns. It could hold two people comfortably, or three if they squeezed in. The driver's seat was at the back of the carriage, behind and above the passengers, who communicated with the driver and paid their fares through a small ceiling hatch. The driver controlled the opening and closing of the cab door with a lever, and could keep the door closed until a passenger paid the fare.[4]

The Clarence was a larger carriage for hire, good for groups of more than two people or for travelers with luggage. The Clarence was commonly known as the "growler" because of the noise its wheels made on cobblestone streets.[5]

Some horse-drawn cab drivers owned their rigs, but most rented them. Besides the daily rent, cabbies had to pay for oil for the carriage lamps and decorative flowers and bells for the horses. It took a generous tip to the groomer to get a well-groomed horse, and another to the washer to be sure the cab would be clean. Drivers also had to pay for their horses to drink water at the cab stand and for permission to wait for fares at railway stations.[6]

In 1891, a German inventor, Wilhelm Bruhn, invented the *taximeter* to measure the distance a vehicle travels. This new device gave drivers an accurate way to determine fares. The name "taxicab" comes from Bruhn's instrument.[7]

A Daimler cab in Ireland, 1899.

By the 1890s, automobiles began to compete with horse-drawn cabs. The Daimler Victoria, built by Gottlieb Daimler in 1897, was the first car built specifically for use as a taxi and came with a taximeter

Yellow cabs in New York City.

Sunny_008/Thinkstock

installed. Daimler sold the car to Friedrich Greiner, an entrepreneur in Stuttgart, Germany, who started the world's first motorized taxicab company.[8] The first motorized London taxi was the electric-powered 1897 Bersey.[9]

By 1899, there were almost a hundred motorized taxis on New York City streets, all of them electric. On September 13 of that year, an out-of-control taxi hit and killed 68-year-old Henry Bliss, making him the first American to die in a car accident. In 1907, Harry Allen imported 600 gas-powered automobiles from France to start the New York Taxi Company,[10] and other taxi fleets followed.

As more taxis appeared on the streets, cities began developing and enforcing regulations for taxi and driver licensing, fares, and pick-up locations. Municipal governments sought to regulate the quality of their taxi services by limiting the number of licenses.[11]

The life of a cab driver today is not all that different from that of an eighteenth-century hackney cab driver. Most cabbies work long hours, and many lease their cabs rather than own them. As of May 2012, roughly half of all cab drivers in the U.S. made less than $10.97 an hour, or $22,820 a year.[12] Bad weather, difficult customers, and heavy traffic make driving stressful, and a good driver still requires intimate knowledge of local streets and landmarks and a good sense of direction.

Ride-Sharing

New ride-share companies like Uber and Lyft are creating competition for taxicab companies. A smartphone application connects people needing a ride with nearby drivers willing to provide one. Passengers pay in advance by credit card online, and no tipping is expected.

Cab companies complain that the competition is unfair because ride-share companies aren't subject to the same regulations and license fees as taxis. (In Boston, for example, every taxi must display a medallion; 1,825 medallions have been issued by the city, and the average price for one was $700,000 in 2014.) The ride-share companies argue that they provide faster, better service, are more affordable, and serve neighborhoods that taxicab companies often avoid. Many cities are struggling to resolve this dispute.

Submarines
To the Bottom of the Sea

Jules Verne's 1870 science fiction novel *Twenty Thousand Leagues Under the Sea* chronicles the hunt for a mysterious sea monster. When the monster damages the vessel that has been hunting it, the protagonists are hurled onto the creature itself, which turns out to be a submarine, the *Nautilus*, commanded by Captain Nemo. Built in secret and powered by an electric motor, the *Nautilus* roams the sea conducting advanced marine research. In the course of the novel, the submarine travels to the Red Sea, the Antarctic ice shelf, and the mythical submerged civilization of Atlantis, surviving an attack by a giant squid and entrapment by an enormous whirlpool. Eventually the three main characters escape to an island, but readers never learn the fate of the *Nautilus* and its captain.

A boat that could travel underwater might have seemed a far-fetched fantasy in 1870, but an Englishman named William Bourne published a design for a submersible boat in 1573. (The boat's mast would have operated as a snorkel to bring in air—but Bourne probably never built it.) In 1620, Cornelius van Drebble, a Dutchman working for the British Royal Navy, built several wooden submarines that were propelled by oars and equipped with tubes to bring in air from the surface. Reports indicate that van Drebble demonstrated his submarines' ability to travel up and down the Thames River[1] at depths of 12 to 15 feet (4 to 5 meters).[2]

NOAA

David Bushnell's *Turtle*.

By 1727, at least 14 patents for submarine designs had been issued in England, but all were essentially submersible rowboats. It wasn't until the American Revolution in 1776 that Yale University student David Bushnell built the *Turtle*, the first propeller-driven submarine. Bushnell designed the *Turtle* to attach explosives to the hull of a ship undetected. The one-person sub was powered by hand-cranked propellers. To submerge, the operator filled a tank at the bottom of the vessel with water; to resurface, he emptied the tank with a hand pump. While the *Turtle* was never used successfully, its design anticipated several aspects of modern submarines.

On February 17, 1864, during the U.S. Civil War, the Confederate sub *Hunley* became the first submarine to sink an enemy ship when it rammed its spar torpedo into the hull of the USS *Housatonic*. Unfortunately for the *Hunley*, the blast destroyed it, too.[3] There would be no other successful submarine attack until World War I, 50 years later.

Engineers had to overcome many challenges to design modern submarines. An effective submarine must be self-propelled with an engine that does not emit exhaust when operating underwater. It must withstand the pressure of the surrounding water and navigate without surfacing. It must be self-contained, providing living quarters, food, and, most importantly, clean air for its crew. Designers have incorporated new technologies over the years to meet these challenges, including advanced propulsion systems, sonar for navigation, and the ability to extract oxygen from seawater via electrolysis.

Modern subs range from two-person vehicles that can operate only a few hours at a time to large, nuclear-powered vessels that can remain submerged for up to six months. In addition to their military uses, submarines are used for scientific research, salvage operations, and exploration. Modified subs are used to repair sea-bottom cables, conduct underwater archaeological missions, and provide tourists with a peek into the beauties and mysteries of the undersea world. In the early 1960s, Donald Reid designed and built the first single-seat craft

A depiction of the Confederate submarine *H.L. Hunley*.

American Civil War Museum

A sailor throws a heaving line to the Ohio-class guided-missile submarine USS *Michigan* as it arrives at Naval Magazine Indian Island, June 8, 2011.

Lt. Ed Early

capable of both flight and underwater propulsion,[4] and as of 2010, the U.S. Defense Advanced Research Projects Agency (DARPA) was developing a flying submarine that can travel at high speeds both over and under the water.[5]

Personal Submarines

Steve Jurvetson

The DeepFlight Super Falcon personal submarine.

With enough money—at least $1.5 million—you can own your own submarine. These small subs are designed to carry one to eight passengers and can operate as deep as 1,000 feet (305 meters). They range from 16 to 21 feet (5 to 6.4 meters) long, with the larger models weighing about 2 tons.[6,7] Some require a trained operator, but others are designed for amateur operation. Manufacturers hope that megayacht owners will want to add undersea exploration to the many amenities their yachts provide.

Stagecoaches
Three Centuries of Service

Royal Coach; Bath to Exeter Stage; Manchester to London Coach, a painting by John Charles Maggs (1819–1896).

Bonhams

We associate stagecoaches with the American West, but they date back to the 1600s in England. Stage-wagons ran regularly between London and Liverpool by 1650, with trips departing every Monday and Thursday and taking 10 to 12 days, depending on weather. By 1698, the early stage-wagons—little more than uncovered horse-drawn wagons—had evolved into enclosed coaches running regular routes on England's three principal roads.[1]

Trish Steel

Sign for the Coach and Horses, a common pub name in all parts of Britain since the seventeenth century. Inns were natural stopping places, providing refreshment for horses and humans.

Long routes were traveled by stages—thus the name—with 10-minute stops every 10 to 12 miles (16 to 19 km) to exchange tired horses for a fresh team and give passengers a chance to stretch. Every 40 to 50 miles (64 to 80 km, a full day or night's travel), the coach would make a longer stop allowing passengers to purchase a meal or possibly a bed for the night. In 1763, the trip from London to Edinburgh, Scotland, took a full two weeks. The coach departed once a month and spent Sunday at an inn to let passengers attend church services and get some much-needed rest.[2]

By the late 1700s, stagecoaches were in regular use in the eastern and southern U.S., where well-established cities and towns provided convenient stopping places. Travel across the largely unsettled American West was much more difficult and dangerous.

In 1858, John Butterfield set up a stagecoach route between St. Louis, Missouri, and San Francisco, California. The nearly 2,800-mile (4,506-km) route ran through El Paso, Texas, and took approximately three weeks to travel in good weather. Butterfield's incentive was a contract to carry the U.S. mail. Because there were no towns along many sections of the route, he set up way stations where drivers could change horses and passengers could get a meal. Many of these stations later grew into settlements.[3]

Early 1861 marked the beginning of the U.S. Civil War, and Texas joined six southern states in seceding from the Union. To avoid the conflict, Butterfield moved his overland route north, across the future states of Wyoming, Utah, and Nevada. The Holladay Overland Mail, started by Ben Holladay, set up an even more ambitious service. Holladay's coaches moved mail and passengers from Missouri River steamboat landings in Kansas and Nebraska through Salt Lake City, Utah, and on to Montana and Oregon.

Stagecoach travel was far from comfortable. The Concord coach, built in Concord, New Hampshire, was considered the best available. The inside of the coach was about four feet wide and four and a half feet high. Three hard, leather-covered benches provided seats for nine passengers. People sitting on the middle bench had no backrest; if they

Library of Congress

"The Deadwood Coach" shows us a side view of a stagecoach with formally dressed passengers.

were lucky there was a leather strap running behind them. The windows were covered with leather curtains that kept out light but not dust, wind, rain, and snow.[4] Stagecoaches that carried the mail traveled day and night, so passengers were obliged to sleep in their seats. Leaning on another passenger's shoulder while sleeping was frowned upon. Leather "boots" at the front and back of the stagecoach carried mail and

Carl Rakeman painting, Federal Highway Administration

The meeting of the east and west branches of the Transcontinental
Railroad in 1869 heralded the end of the stagecoach business.

passengers' luggage. When the boots were full, bags of mail would be loaded around passengers' feet.

Many stagecoach routes were little more than narrow tracks over rugged terrain. Deep sand and mud routinely covered portions of the trail. One selling point of the Concord Coach was its ability to keep passengers dry while floating across streams.[5] On steep hills, passengers were asked to get out and walk to lighten the load, and occasionally to push as well. Mail coaches hired an armed guard to ride with the driver to prevent robberies.

The growth of railroads, and especially the completion of the Transcontinental Railroad, heralded the end of the stagecoach business. Some stagecoaches continued to serve remote towns off the rail lines until automobiles and buses came into use, but by the beginning of World War I, the age of the stagecoach had ended.[6]

Buses
Another of Pascal's Great Ideas

In 1662, Blaise Pascal—the famous mathematician and philosopher—started the first known horse-drawn public bus service in Paris. Pascal began with seven coaches that carried six to eight passengers each along regular routes. The service was popular at first but went out of business by 1675.[1] Not until the early 1800s did a public bus service reappear.

Early horse-drawn buses were called *omnibuses*—from the Latin "for all"—because anyone who could pay the fare could ride them. Jacques Lafitte started an omnibus company in Paris in 1819, and five years later, across the English Channel, John Greenwood began a service carrying merchants and shoppers to Market Street in Manchester, England.[2] In 1826, Stanislas Baudry initiated omnibus service from Nantes, France, to a bathhouse he had built outside the city. Upon realizing that most riders were not using the bathhouse, he closed it to concentrate on the bus system.[3] Within two years, he moved to Paris and started the General Omnibus Company.

George Shillibeer, a British coach builder working in Paris, thought London should have these omnibuses, and in 1829, he started the first London service. The hackney-cab companies, who had enjoyed a monopoly until then, tried to block Shillibeer's new business, and his omnibuses were not allowed to operate in the city center until 1832.[4]

The Clapham to Putney, England, omnibus in 1866.

While hackney-cab passengers had to arrange for a pick-up or go to a cab stand, omnibuses followed a set route on a fixed timetable, and passengers simply waited along the route for the next bus to come by. Omnibus fares were cheaper than cabs.

On the other hand, omnibuses were slow and uncomfortable. Traffic congestion made it hard for drivers to stay on schedule, but they could be fined for driving too fast. Most omnibuses were licensed to carry 10 to 12 passengers inside and another 10 to 14 outside, but conductors often squeezed in an extra passenger or two and kept the extra fares for themselves. Before companies began selling tickets in 1891, some conductors would quote one price when a passenger boarded and a higher price when collecting the fare, lining their pockets with the difference. Unlicensed "pirate" omnibuses charged illegal higher fares, and some even carried a pickpocket on board, who paid the driver a cut of the stolen goods.[5] Despite such hazards, omnibuses provided affordable transportation, and demand for service continued to grow.

By the 1830s, steam-powered buses and electric trolley buses powered by an overhead cable had begun to compete with horse-drawn omnibuses. The first gasoline-powered bus began service in 1895, and buses evolved in tandem with automobiles throughout the 1900s. They comprise an important part of a city's transportation network and are relied on by many city dwellers. Long-distance bus companies link major cities.

Modern buses are comfortable and fuel-efficient. Low-floor designs, developed in the 1990s, make them easier to board, and some have wheelchair ramps or lifts and bicycle racks. Articulated buses— built with multiple compartments that bend in the middle—are longer and can carry more passengers while turning corners safely.[6]

The next generation of buses may be electric. New lithium-ion batteries are stronger than earlier ones and can go longer without recharging, and companies are experimenting with rapid-recharging stations that can be used for a few minutes at a time, such as at a regular bus stop.[7] Vienna, Austria, began running a small fleet of 12 electric buses in 2013. The buses are fully recharged each night, and during the day they can connect to the power grid used by the city's trams and subway trains for 10- to 15-minute partial recharges.[8]

An electric minibus in St. Helens, England.

Jeepneys and Chicken Buses

Not all buses are purpose-built. Jeepneys, the colorful minibuses used in the Philippines, are surplus World War II jeeps that have been stripped down to carry more passengers.

Magalhães

Philippine jeepney at the Carbon Market.

The chicken buses of rural Honduras and Guatemala are old American school buses that have been modified and decorated. They are called chicken buses because passengers often carry chickens or other livestock with them. Pakistani drivers adorn their buses with colorful paintings and decorative trim, each one a unique work of art.[9]

The Yellow School Bus

Josephine Barker Collection, University Archives, University of Kentucky, Special Collections

A horse-drawn school bus packed with students in 1911.

George Shillibeer, who started the first London horse-bus service, also created the first school bus in 1827, carrying 25 students to the London Quaker School. Wayne Works, a U.S. company, began building horse-drawn school buses around 1886. These buses had doors in the rear so that children getting on and off wouldn't startle the horses. By 1910, thirty states had bus transportation programs, and by the 1930s, a variety of motorized vehicles were being used as school buses. A conference in 1939 set 44 safety guidelines for U.S. school buses, leading to standardized models. One required that buses be painted yellow to make them highly visible to other traffic.[10]

Automobiles
"Get a Horse!"

When the first steam-powered automobile appeared on the streets of Racine, Wisconsin, in 1870, it caused a panic. Its loud steam whistle was startling, and people were afraid that the engine would explode. Critics claimed that these new self-propelled vehicles frightened horses and would put untold numbers of carriage drivers, grooms, and blacksmiths out of work. Others said that automobiles were too fast and endangered pedestrians. In 1865, the United Kingdom passed the Locomotive Act, also known as the Red Flag Act, limiting the speed of automobiles to 2 miles (3 km) per hour in the city and requiring that a man waving a red flag walk in front of the vehicle to warn of its approach.[1] Parliament removed the red flag requirement in 1878, but the act itself was not repealed until 1896.

Nicolas-Joseph Cugnot invented the first self-propelled road vehicle in 1769, a steam-powered tractor to pull cannons for the French army. The vehicle had three wheels and a top speed of 2.5 miles (4 km) per hour, and it had to stop every 10 to 15 minutes to build enough steam pressure to resume moving. In 1770, Cugnot built a steam-powered four-passenger tricycle, and in 1771, he suffered the first documented motor vehicle accident when he drove one of his inventions into a stone wall.[2]

Steam engines worked by heating water in a boiler, creating steam that expanded in cylinders to push pistons, which then turned a vehicle's

Joe deSousa

Cugnot's steam-powered vehicle of 1770 is displayed at the
Conservatoire National des Arts et Métiers in Paris.

wheels via a crankshaft. The boiler and its firebox—where wood or coal
was burned—were external to the engine; picture a railroad engineer
shoveling coal into the firebox of a boiler car behind the locomotive.
Steam had to be piped under pressure from the boiler to the engine cylinders, with attendant loss of heat and efficiency. Cugnot's early vehicles
weighed an estimated 8,000 pounds (3,629 kg) and were only practical on flat, hard roads or iron rails.[3] A practical automobile required a
lighter but still powerful engine.

Inventors began wrestling with the design of a workable *internal combustion engine*—in which combustion would take place inside the cylinders—in the late 1600s. Christian Huygens, a Dutch chemist, designed
an engine in 1680 that used gunpowder as a fuel. In 1807, Francois
Isaac de Rivaz of Switzerland created an engine that burned a mixture

of hydrogen and oxygen. Other inventors tried coal gas, kerosene, and petroleum with varying success.[4]

In 1885, Gotlieb Daimler patented the engine that is generally considered the modern prototype, with gasoline injected into the cylinder through a carburetor. A year later, he installed his engine in a converted stagecoach, creating the first four-wheeled gas-powered automobile. That same year, Karl Benz received the first patent for a gas-fueled car, and in 1890, Wilhelm Maybach built the first four-cylinder, four-stroke engine. Maybach later designed the Mercedes.[5]

Gasoline cars were outselling all other motor vehicles by the early 1900s. In November 1900, when New York's Madison Square Garden hosted the first U.S. National Automobile Show, 31 exhibitors show-

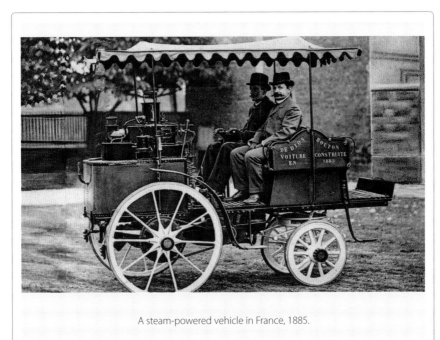

A steam-powered vehicle in France, 1885.

De Dion-Bouton

cased their products.[6] But car manufacturers still needed to convince people that automobiles were safe and reliable; cries of "Get a horse!" often greeted early adopters. In 1900, the Automobile Club of Great

Britain and Ireland held the 1,000 Miles Trial, in which 65 cars left Hyde Park Corner in London for a tour of major English and Scottish cities. Most of the cars finished the tour successfully.[7] In the U.S., manufacturers staged similar "reliability tours" to familiarize people with this new vehicle.[8]

Mass production and lower prices helped fuel the automobile's growing popularity. In 1913, more than 600,000 automobiles were produced worldwide.[9] Streetcars, buses, and other forms of public transportation had freed people from having to live close to their jobs, and the automobile continued this trend. Commuters began to leave cities for suburbs. New highways facilitated automobile travel, and inventors continued to introduce new technologies to improve the comfort, safety, and efficiency of automobiles.

Henry Ford and Mass Production

Henry Ford is often the first name that comes to mind when talking about the history of the automobile. Ford's genius was his desire to build a car that would be inexpensive, reliable, and easy to drive. His Model N, released in 1906, was hailed as the first low-cost car that was "well built and offered in large numbers."[10] The Model T, which

The assembly line at the Ford Motor Highland Park plant in 1923.

followed in 1908, was even more popular: 5,896 were sold the first year at $850 each, and in 1912, 78,611 were sold at $600 each.[11] To meet the growing demand, Ford invented a moving assembly line that reduced the time required to build a car, and in so doing revolutionized manufacturing.

The Automobile and Demand for Oil

A Model T Ford at the Doulton Fountain on Glasgow
Green, Glasgow, Scotland.

By the mid-1800s, whale oil was being used to fuel lamps throughout colonial America and elsewhere, but whale oil was expensive. Camphene, a mixture of turpentine and alcohol, was cheaper but highly flammable, and often caused house fires.

In 1846, Canadian Abraham Gesner developed a process to refine a liquid fuel from coal or oil. It was cheaper and cleaner than whale oil and safer than Camphene. Gesner called his new fuel kerosene, but consumers often called it coal oil. (Kerosene is only slightly more refined than today's home heating oil, and in fact is used in the furnaces of mobile homes with outside storage tanks in northern climates where ordinary heating oil might gel in midwinter.) By 1860, more than 30 companies were producing kerosene, and the U.S. Patent Office recorded almost 250 patents for lamps, wicks, burners, and fuels.[12] There was only one problem: refining oil to make kerosene created a useless byproduct—gasoline.

By 1910, automobiles were becoming more common, and inventors knew that internal combustion engines ran best on light fuels like gasoline. Suddenly gasoline was no longer a waste product. Now the problem was how to produce enough of it.

(Continued on next page)

(Continued from previous page)

William Burton and Robert Humphreys, chemical engineers at Standard Oil of Indiana, had tried unsuccessfully to change the distilling process to produce more gasoline from a given volume of crude oil. Finally, Burton suggested adding pressure to the standard heating process, thus causing the heavier molecules of kerosene to "crack" into lighter molecules such as those of gasoline. Called *thermal cracking*, the process doubled refining efficiency. In 1930, a Frenchman named Eugene Houdry further refined the cracking process, making it even more efficient.[13]

Mass production, lower prices, and new and better highways continued to increase demand for automobiles and the gasoline to run them. Today, the U.S. uses 28 percent of its energy for transportation. More than 60 percent of U.S. transportation energy is used for personal vehicles, and 86 percent of that energy comes directly from gasoline and diesel fuels.[14]

Hot Air Balloons
Up, Up, and Away

The three passengers on the first "manned" hot air balloon flight were a sheep, a duck, and a rooster. Launched on September 19, 1783, at Paris, France, the balloon flew 10 to 15 minutes before crashing to the ground.[1] The animals were unhurt, and the experiment was considered a success, proving that it was possible to build a balloon large and strong enough to carry passengers.

Unmanned hot air balloons had been around a long time, having been used in China as airborne lanterns for military signaling as early as AD 220.[2] Hot air balloons could have been used in creating the famous Nazca Lines in Peru between AD 400 and 650.[3] But using balloons for human flight was a new idea in 1783.

Experiments with human passengers followed that first animal flight. At first, the balloon was tethered to the ground and ascended no higher than the length of its tether. In November 1783, Pilatre De Rozier became the first person to ascend untethered. His balloon reached a height of 500 feet (152 meters) and flew approximately five and a half miles (9 km) before landing 20 minutes later.[4]

Hot air balloons work on the principle that warm air rises through the cooler, denser air around it. The "balloon" is called the *envelope*, and a large basket (traditionally wicker) called the *gondola* hangs from it to carry passengers. The pilot fires up a propane burner above the gondola, in

In 1878, 26 years after making the first successful flight in a self-propelled airship (Chapter 29), Henri Giffard offered Parisians their first-ever chance to rise above the city's rooftops in a tethered hot air balloon. Hundreds of men in the courtyard of Paris's Tuileries Gardens operated the system of winch drums and cables that restrained the huge balloon while it was aloft and ultimately reeled it back to earth.

the center of the envelope's opening, to further heat the air in the envelope, causing the balloon to ascend. Venting hot air from the top of the envelope causes the balloon to descend. The balloon's direction of travel, however, is harder to control. It follows the prevailing wind, and the pilot can only ascend or descend in search of wind from the desired direction.

In 1785, French balloonist Jean-Pierre Blanchard and American John Jeffries completed the first successful flight across the English Channel. George Washington was on hand for the launch of the first flight in North America in 1793, a 45-minute trip from Philadelphia to Gloucester County, New Jersey. In 1932, Auguste Piccard, a Swiss scientist, used a balloon with a pressurized aluminum gondola to fly into the stratosphere, reaching an altitude of more than 52,000 feet (10 miles or 15.8 km).[5] Today's altitude record of 113,739.9 feet (21.5 miles or 34.7 km), reached in 1961 by Commander Malcolm Ross and Lieutenant Commander Victor Prather of the U.S. Navy,[6] is three times higher than the average altitude of a passenger plane. The past 40 years have brought the first crossings of the Atlantic (1987) and Pacific (1991) oceans and the first successful flight around the world (1999), which took 19 days, 21 hours, and 55 minutes to complete.

But hot air balloons have never been considered a practical means of transportation. Balloons can travel only as fast as the wind blows, and the winds can be unpredictable and shifting, so the pilot cannot accurately predict where a flight will land. Rarely is it possible to land where a balloon took off. Finally, hot air balloons fly best in clear skies and light winds. High winds, poor visibility, and possible lightning strikes would preclude a launch.

An advertisement in *Moving Picture World* for the American film *The Shielding Shadow* (1916).

Today, most hot air ballooning is recreational. You can book a balloon ride from a commercial vendor to celebrate a special occasion or view the world from a different perspective. Hot air balloon festivals are popular in countries throughout

Eric Ward

Hot air balloons at play.

the world. A sunrise hot air balloon ride over Tanzania's Serengeti National Park can be a memorable part of an African wildlife safari.

Nazca Lines

Aerial photo of a dog image in the Nazca Lines.

The Nazca (or Nasca) Lines in the desert of southern Peru were formed as long as 2,000 years ago by removing topsoil to reveal pale subsurface sediments that contrast with the darker surrounding surface rocks and sand. Only when viewed from the air do the lines form enormous patterns. There are more than 300 geometric figures, 70 animal and plant designs, and 800 straight lines, some up to 30 miles (48 km) long. Some of the animal and plant designs are as large as the Empire State Building. The designs appear to date from the first seven centuries AD.[7]

No one knows who made the Nazca lines or why. Were they an astronomical calendar, a map of underground water sources, a place to perform important rituals? It has even been suggested that they marked a landing strip for aliens. But decorations on Nazcan pottery closely resemble the images the lines create when viewed from the air.[8] This and the enormous size of the images suggest that the original designers had a way to look down on earth from a great height.

Steamships
Ocean Travel on a Schedule

The Industrial Revolution brought enormous changes to the Western world in the eighteenth and nineteenth centuries. New inventions automated tasks such as spinning and weaving that had formerly been done by hand in people's homes. Production moved to factories, and productivity increased. Cheaper methods of casting iron and steel led to greater use of these materials, and the steam engine provided a new source of power.[1]

The first attempts to power ships with steam engines were not entirely successful, however. Early steamships used coal-fired engines to turn a paddle wheel, but the mechanisms for converting the up-and-down motion of a piston into the circular motion of a paddle wheel were crude. Another problem was building a boiler big enough to create the necessary power[2] and strong enough to withstand the pressures generated by expanding steam. Boiler explosions were common. Eventually these and other problems were overcome.

The first successful steamship was the *Charlotte Dundas*, built in 1802 by William Symington for use as a tugboat on the Forth-Clyde canal in Scotland. American inventor Robert Fulton saw the *Charlotte Dundas* and was inspired to develop his own steamships, improving the design with each one.[3]

The *Great Western* riding a tidal wave, December 11, 1844.

National Maritime Museum, Greenwich, London

Early steamships were used on rivers, lakes, and canals. As steam-ship traffic increased, agencies such as the U.S. Army Corps of Engineers worked to clear rivers of tree trunks, sunken vessels, and other obstacles.[4] In 1825, the state of New York completed work on the Erie Canal, linking the Hudson River to Lake Erie and providing access to the Great Lakes from the Atlantic Ocean.[5] Canadian construction of the Lachine and Welland Canals made it possible for ships to sail between the St. Lawrence River and Lake Ontario, and from there south to Lake Erie. With each improvement, travel times fell.

Until the early 1800s, ocean crossings were made under sail and could take two months or more, depending on wind and weather.

Passengers referred to the sailing vessels carrying immigrants to America as "coffin ships" because of the dangers.[6] In 1819, the paddle wheeler *Savannah* became the first steamship to cross the North Atlantic when it sailed from the Savannah River in Georgia to Liverpool, England, in 27 days. Like many early steamships, the *Savannah* was outfitted with sails to take advantage of favorable winds, minimize fuel consumption, and provide a backup if the engine failed. The *Savannah* used its engine only 85 hours during the entire trip.[7] It would be more than 25 years before a ship powered only by steam would make the crossing, and then it would be a race.

In 1837–38, the paddle-driven steamer *Great Western* was built in Bristol, England, specifically to cross the Atlantic under steam alone. Determined to steal the limelight, an American, Julius Smith, chartered another ship, the *Sirius*, to race the *Great Western* across. The *Sirius* arrived in New York 18 days after leaving Cork, Ireland, but only after the crew had burned all of the coal on board and thrown pieces of the ship's wooden superstructure into the boiler's firebox. The *Great Western* won the publicity contest, leaving Britain four days after the *Sirius* and arriving in New York only hours behind with 200 tons of coal remaining and her structure intact.[8] The *Great Western* completed 45 Atlantic voyages to and from New York before being sold to the Royal Mail Steam Packet Company for service between Britain and the West Indies.

Paddle-wheel steamers were not well suited to ocean crossings, however. In rough seas, a paddle wheel could become completely submerged—overloading the engine—or rise out of the water, unloading the engine and allowing it to race. By the 1840s, shipbuilders were replacing paddle wheels with more efficient screw propellers. Inventors developed more powerful and efficient engines that burned less coal and could power larger ships, and shipbuilders began using iron and steel for hulls.[9] By the late 1800s, steamships had become such a fast, reliable method of travel that Phileas Fogg, the main character in Jules Verne's novel *Around the World in 80 Days*, made more than half his globe-girdling trip by steamship.

Working Paddle Wheelers

Ruhrfisch

The *Hiawatha*, a paddle wheeler on the West Branch of the Susquehanna
River at Susquehanna State Park, Williamsport, Pennsylvania.

Today, when people think of paddle wheelers, they think of the large, fancy
showboats that provided floating entertainment on the Ohio and Mississippi Rivers. Showboats were colorfully decorated and contained theaters, restaurants,
and ballrooms for plays, concerts, and vaudeville. The arrival of a showboat was
cause for excitement in riverside towns. In reality, however, most paddle wheelers were smaller working riverboats that towed barges, ferried people, and transported mail and freight.[10]

Steam Locomotives
All Aboard!

The first railways were short lines connecting a mine or quarry to a ship on a nearby river or canal, with horses pulling a single car along wooden tracks. In 1803, the horse-drawn Surry Iron Railway started carrying freight between Croydon and Wandsworth, England. Four years later, when the first passenger railway, the Swansea and Mumbles, began operations, it too was horse-powered.[1]

On February 21, 1804, Richard Trevithick's high-pressure steam locomotive pulled 10 tons of iron, 70 passengers, and five wagons for 9 miles (14.5 km) while reaching a speed of almost 5 miles (8 km) per hour.[2] The use of high-pressure steam made Trevithick's engine lighter and more fuel-efficient than earlier engines.[3]

Steam locomotives could pull heavier loads for longer distances than horses, and railways could bypass the congestion on local roads and reach areas without natural waterways. Many early steam engines were too heavy, however, and damaged the tracks. Some mine owners preferred to use a stationary steam engine to wind a rope that pulled carts along the tracks. The Stockton & Darlington Railway, which opened in Britain in 1825, was designed to use a combination of steam locomotion, horse power, and rope haulage with stationary engines on iron rails. It wasn't until the 1830 opening of the Liverpool & Manchester Railway that a public railway was built with "all the basic characteristics of a modern rail system: a double track of iron rails used to transport

Catch me who can .

Mechanical Power Subduing Animal Speed .

A drawing of Richard Trevithick's "Catch Me Who Can" locomotive, from a card or admission ticket to the "Steam Circus," summer 1808.

passengers and freight; complete steam locomotive drive; a system of stations, bridges, and tunnels; and even three different classes of carriages."[4]

Passenger comfort was not a big concern on early trains. The first passenger cars were open wagons, and riders had to endure foul weather as well as soot and sparks flying back from the locomotive. Later, builders added roofs, followed by wooden benches and then cushioned seats.

Steam locomotives needed regular resupplies of water for their boilers and wood or coal for their fireboxes. Special train cars carried extra coal and water, and railway stations featured large water towers so that trains could refill their tanks. High boiler pressures made explosions a constant threat. After one boiler explosion, the South Carolina Canal & Railroad announced that it would put a car loaded with bales of cotton between the engine and the passenger cars to protect travelers.[5]

Railways around the globe connected cities and opened remote areas for settlement. The Transcontinental Railroad, which enabled passengers to travel coast-to-coast in the U.S., was completed in 1869. The Canadian Pacific Railroad, the Alaska Railroad, and the Trans-Siberian Railway, which traveled across Russia to China, were all built to provide access to natural resources in areas that had previously been inaccessible or ignored. Migrations and settlements followed the rail lines.

As railway trips grew longer, an emphasis on passenger comfort took hold. In 1859, George Pullman, an American, designed a sleeping car and started an era of luxury rail travel.[6] The Orient Express, put into service in 1883 between Paris, France, and Istanbul, Turkey, was one of

The Valley Railroad steam engine in Essex, Connecticut, December 18, 2004.

the best-known luxury trains. Sleeping compartments featured silk sheets, marble sinks, and gold plumbing fixtures. Passengers dressed formally to eat in a dining car with fine china and crystal chandeliers.[7]

The rise of the internal combustion engine doomed the steam locomotive. Between 1940 and 1955, the number of steam engines on American railroads dropped from 40,000 to 2,000, while the number of diesel engines rose just as dramatically. By 1955, diesel locomotives

How the U.S. Got Its Time Zones

The completion of the Transcontinental Railroad in 1869 highlighted the need for a standardized system of timekeeping. With 60 time zones across the U.S., train schedules for passengers were very confusing. Passengers could leave one station at noon, travel for two hours, and arrive at their destination at 12:30 p.m. by the local clock. To create a common timetable for scheduling, the railroads developed the current system of four time zones, which became law in 1883.[9]

made up about 90 percent of all rail traffic.[8] Today, only a few steam locomotives remain in operation, mostly for sightseeing excursions or at historic sites established to preserve them.

The Woman in the White Dress

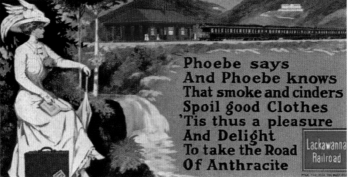

A September 1912 postcard depicting the Lackawanna
Railroad's advertising character, Phoebe Snow.

In 1900, the Delaware, Lackawanna and Western Railroad (DL&W) hired a New York advertising agency to help increase passenger traffic, and the result was Phoebe Snow.

Burning coal produces fumes and ash. Train compartments were often hot and stuffy, especially in the summer, but if passengers opened windows, they could be covered in black soot. The solution was anthracite coal. Found mainly in northeastern Pennsylvania, this hard coal burns hotter and produces less ash than soft bituminous coal, and the DL&W used anthracite.

Advertisers decided to promote the idea that DL&W passengers arrived at their destinations looking clean and fresh. Ads pictured Phoebe Snow in her clean white dress boarding and riding the train, eating in the dining car, and enjoying the view from the observation deck. Each ad was accompanied by a short rhyme in which Phoebe proclaimed the benefits of traveling "upon the road of Anthracite."

The ad campaign increased ridership and ran until the beginning of World War I, when the use of anthracite coal was restricted to the war effort. In 1949, the Lackawanna Railroad inaugurated a new streamlined passenger train named the *Phoebe Snow*, which ran until 1996.[10]

Amphibious Vehicles
One If by Land, Two If by Sea?

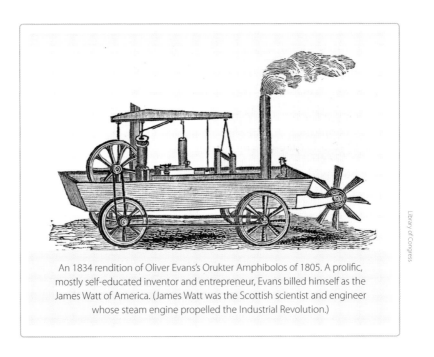

An 1834 rendition of Oliver Evans's Orukter Amphibolos of 1805. A prolific, mostly self-educated inventor and entrepreneur, Evans billed himself as the James Watt of America. (James Watt was the Scottish scientist and engineer whose steam engine propelled the Industrial Revolution.)

Library of Congress

In 1805, when the Philadelphia city council wanted to deepen a section of the Schuylkill River, Oliver Evans, a local inventor, built what he said was the perfect vehicle for the job: the *Orukter Amphibolos* (Greek for "amphibious digger").[1] As Evans described his invention:

> I first put wheels to the boat and propelled it by the engine a
> mile and a half up Market Street and around Center Square
> to the River Schuylkill. I then fixed a paddle-wheel at the

A duck boat on the beach at De Panne, Belgium.

Lokilech

stern and propelled it by the engine down the Schuylkill and up the Delaware sixteen miles leaving all the vessels that were under sail full halfway behind me.[2]

Evans's steam-powered vehicle was so heavy that the wheels and axles broke the first time he tried to move it, and later he damaged the road. The Orukter may never actually have made it into the water—Evans was known to exaggerate—and it was never used again.[3] But it went down in history as the first amphibious vehicle.

In 1849, Mr. Gail Borden, a Texas inventor, built a watertight, sail-powered wagon that ran well on land but tipped over in the water, the hull being too light to stand up to the wind in its sail.[4] In 1883, a Mr. Terry became the first person to cycle across the English Channel, pedaling an amphibious tricycle to the water's edge, taking it apart, reassembling it in a different form, and covering it with stretched canvas to create a rowboat.[5] The Beach and Harris Cycle Raft, invented in 1894, used two inflatable pontoons to float a bicycle. And in 1907, a Frenchman named Ravailler built a canoe-shaped car. Once in the water, the car was driven by a propeller, and the wheels acted as rudders.[6]

The first really useful amphibious vehicle was the Alligator Tug, developed for logging companies in the 1870s to break up river logjams.

The Alligator Tug was a steam-powered paddleboat that used a winch and a large anchor to pull itself out of the water and across land to the next waterway. Although the tugs' top speed on land was only 2 miles (3.2 km) per hour, they worked well in the water.[7]

Light gasoline engines gave new life to the idea of an amphibious vehicle, and by the late 1920s, inventors were combining an automobile chassis with a boat hull and oversized wheels to create an amphibious car. But it was World War II that spurred development of the best-known amphibious vehicle, the DUKW (pronounced *duck*).

The DUKW was developed by the United States Army to land troops on beaches and ferry supplies ashore. Based on the body of a 2.5-ton General Motors cargo truck, the DUKW was designed to perform as well as other trucks on land, handle rough seas, and drive over reefs and sandbars. It proved invaluable in the Allied invasions of Salerno, Normandy, Manila, and Okinawa, as well in many other conflicts.[8] Today, many cities built around rivers and bays use refurbished DUKWs to

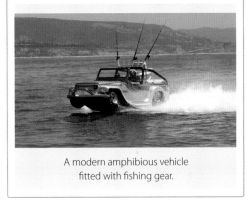

A modern amphibious vehicle fitted with fishing gear.

WaterCar

give "Duck Boat" tours that cruise both land and water.

The Amphicar, built in Germany in 1961, was one of the first amphibious vehicles to be sold for personal use, and inventors continue to develop them. In 2004, Rinspeed, a Swiss company, created a vehicle that converts from a sports car with a top speed of 125 miles (201 km) per hour to a speedboat that can reach 80 miles (129 km) per hour. The 2012 Gibbs Sport Quadski amphibious car claims to be as fast on water as on land.[9] Creative tinkerers have built an amphibious 42-foot (13-meter) motorhome, an amphibious truck made from a stainless steel milk-truck tank, amphibious motorcycles, and an amphibious Lamborghini.[10]

The amphibious bike "Cyclomer" in Paris, 1932. For water travel, the rider lowered the four smaller spheres (two on the handlebars, two on the rear frame) to wheel height.

Around the World in 10 Years

In 1958, Ben Carlin, an Australian, completed his quest to drive around the world in his amphibious vehicle named the *Half-Safe*. The 10-year trip, which started and ended in Montreal, Canada, covered 11,050 miles (17,783 km) by sea and 38,987 miles (62,743 km) on land. The trip was delayed

Ben Carlin's *Half-Safe* in Copenhagen in 1951.

numerous times by breakdowns, bad weather, and the need to raise money to continue. To date, Carlin's trip is the only documented circumnavigation of the earth by an amphibious vehicle.[11]

Covered Wagons
On the Oregon Trail

By the 1830s, the westward migration of pioneers had pushed the American frontier to the Mississippi Valley. Cities such as Cincinnati, Detroit, Chicago, and St. Louis were growing rapidly, and the newly settled Midwest and mid-Southern regions supplied the eastern U.S. with food, timber, minerals, and other resources. But many people still hoped to make their fortunes on new frontiers. Traders, explorers, and missionaries returning from California and Oregon brought tales of fertile valleys with abundant farming, great forests, and hunting and mining opportunities. From the 1840s to the 1860s, more than 300,000 pioneers, most traveling in covered wagons, crossed plains, rivers, deserts, and mountains to reach this new frontier.[1] The route they followed was called the Oregon Trail.

Pioneers faced a daunting task. They had to travel approximately 2,000 miles (3,200 km)—from Independence, Missouri, the start of the trail, to their destination in Oregon or California—carrying with them everything they needed for the trip and a new life.[2] They also had to carry enough tools and spare parts to repair their wagons en route, and they had to complete the trip between the end of one winter and the beginning of the next. Bad weather could be dangerous, especially in the high mountain passes where snow lingered most of the year. And

Beinecke Rare Book and Manuscript Library/Yale University Library

A Currier & Ives lithograph of a wagon train crossing the Rocky Mountains.

they had to be prepared for births, deaths, illnesses, and injuries that could happen anywhere along their journey.[3]

The covered wagon answered many of these problems. It was large enough to hold household goods and supplies, provided shelter from the weather, and could travel over rough terrain. Often a converted farm wagon, its bed was a rectangular wooden box, approximately ten feet by four feet, with two-foot-high wooden sides. On some wagons, the front and back ends curved upward to keep items inside from rolling out.[4]

Pioneers packed heavy supplies such as a plow, a bag of seed, a stove, or a spinning wheel in the bottom of the wagon to keep the center of gravity low for stability. Lighter goods, such as kitchen utensils and clothes, made up the next layer. Blankets, food, cooking pots, and other necessities for the trip were stored on top, and everything was carefully fitted and strapped down to keep the weight from shifting.[5]

Travelers had to bring everything they needed without making the wagon too heavy to pull. Often, as the trail became steeper and the animals pulling it grew more tired, goods were abandoned to lighten the load.

Five or six large hoops of bent hickory stretched from one side of the wagon bed to the other supported a waterproof canvas cover. Flaps at the front and a "puckering string" that closed up the back could be opened for ventilation or closed to protect against bad weather. These large white canopies "sailing" across the prairie helped give covered wagons the nickname "prairie schooners."

The undercarriage included the wheels, axles, and a long tongue for yoking the animals, usually oxen, that pulled the wagon. The wagon's front wheels were smaller in diameter than the rear wheels to make turning easier. Some wagons had a long lever on one side that the driver could pull to push brake pads against the wheels; others had no brakes at all. Travelers often used chains or small trees wedged into the rear wheels to provide braking on steep downslopes.[6]

Busting the Myths about Covered Wagons

Most of what old TV shows and movies show about wagon trains isn't true. The large Conestoga wagons they show were used for hauling freight over roads, but were too large and heavy for the Oregon Trail. Horses could be used to pull the wagons but were expensive to purchase and required large quantities of grain for feed. Mules were strong and fast but could be stubborn, skittish, and hard to work with. Oxen were the strongest, cheapest, and most docile, and they could graze along the trail, making them the best choice for the trip.[7] The driver didn't ride, but walked alongside the wagon, guiding the oxen with verbal commands and whip cracks. Nobody else rode either, unless they were unable to walk. For the most part, families walked along the trail—15 to 20 miles (24 to 32 km) a day—and slept outside at night.[8] Indian attacks were rare, and most contact with Native Americans was for trade. If pioneers circled their wagons, it was more likely as a defense against wolves or other predators.[9]

An Oregon Trail reenactment.

For safety and mutual assistance, pioneers traveled together in wagon trains. Trains of 20 to 30 wagons were common, but some were many times larger. With cows, sheep, and extra oxen rounding out the procession, a wagon train became a moving small town.

Newest Isn't Always Best

By 1840, people were traveling by stagecoach, steam trains, and steam-powered buses and trolleys. So why did the westward migration in the U.S. rely on ox-drawn covered wagons and not these more modern technologies? The answer is simple: covered wagons were better suited to the task.

Stagecoaches and buses needed roads. Steam locomotives needed railroad tracks. And steam-powered vehicles of any kind needed regular resupplies of wood or coal to keep their boilers working. The Oregon Trail had none of these things.

Covered wagons provided shelter from the elements and the storage space that pioneers needed to carry their belongings. The wagons were sturdy enough to travel over unpaved grasslands and rocky terrain. The oxen or other animals pulling the wagons could eat available vegetation. With few moving parts, wagons could be repaired along the trail. The Oregon Trail provides a good reminder that the best technology for a given task may not be the newest one.

Ocean Liners
Huddled Masses in Steerage

I n popular culture, the words "ocean liner" evoke the wealth, luxury, and glamour on display in the movie *Titanic*. But in reality, most passengers on ocean liners were poor emigrants seeking work or fleeing upheaval and oppression in their home countries, and their shipboard accommodations were far from luxurious.[1]

By the mid-1800s, steamships had proven themselves safer, more reliable, and more comfortable than sailing ships. As people became more willing to travel by sea, shipping companies found that passenger fares could be a lucrative adjunct to cargo and mail,[2] and they began to pay more attention to making passenger compartments comfortable and attractive.

From the mid-1880s until the late 1960s, ocean liners were the primary means of travel between continents. Between 1880 and 1930, more than 27 million people immigrated to the U.S. from countries around the world,[3] the majority by ship. Shipping lines made far more money from transporting emigrants than from catering to wealthy passengers.

First-class passengers paid the highest fares for the most luxurious quarters. Their staterooms compared favorably with rooms in the best hotels, and they enjoyed their own dining and sitting rooms. Second-class passengers received less luxurious accommodations, and passengers in third class, or *steerage*—the lower deck or decks where the majority of

HITE STAR LINE R.M.S. "BRITANNIC" 50,000 TONS.
LAUNCHED FEB. 26TH 1914

Postcard of the RMS *Britannic*, sister ship to the *Titanic*.

emigrants traveled—had only the barest amenities. As one example, the deck plan of the SS *Japan*, which transported Asian emigrants across the Pacific to California, could accommodate 190 first-class passengers in 50 staterooms, whereas the 908 passengers who spent the month-long voyage in steerage slept in rows of open bunks stacked three high in the space closest to the engine room. And the *Japan*, like other liners, often carried more passengers than legally allowed; in 1873, her captain was cited for carrying 451 additional passengers on a single voyage.[4]

Ocean liners played an important role in both world wars of the twentieth century. Large liners were pressed into service for troop transport and hospital ships, while smaller ones became armed merchant cruisers. Some became "Q-ships," bait to lure and entrap enemy submarines. The RMS *Britannic*, sister ship to the *Titanic*, was built as a luxury ocean liner but was commissioned instead as a World War I hospital ship, and was sunk by a mine a year after going into service.[5] Many liners, even those not used for military purposes, became targets for German U-boats and enemy aircraft. The sinking of the RMS *Lusitania* in 1915 is widely credited for America's entry into World War I in 1917. Of the 1,198 passengers killed, 128 were Americans.[6]

It wasn't until 1958 that an airline, Pan American World Airways, instituted regular flights between New York and Paris,[7] a trip that took

Titanic first-class stateroom B-59, decorated in Old Dutch style.

National Archives

A third-class cabin in the *Titanic*.

seven hours by air versus seven to eight days by sea. As air travel became safe and more affordable, passenger traffic on ocean liners declined, though cruise ships (Chapter 35) have flourished. Today, the only ocean liner still in service is the *Queen Mary 2*, built in 2003 and used both as an ocean liner and for cruises.[8] At 1,132 feet (345 meters) long, she's the longest passenger ship in the world and can carry more than 2,600 passengers with a crew of 1,200. Four other liners have been preserved as museums or floating hotels and restaurants.

Dirigibles
Self-Propelled Gas Bags

Hot air balloons (Chapter 23) enabled people to fly, but only where the wind took them. Ideas for controlling the speed and direction of flight included outfitting balloons with flapping wings; oars and a tiller to row through the air; or propellers. Some inventors built detailed models of their sometimes whimsical airships, which were called *dirigible balloons*, French for "steerable balloons." Later this was shortened to *dirigibles*.

In 1852, Henri Giffard flew the first successful full-scale airship from the Paris Hippodrome to Trappes, a distance of 17 miles (27 km). His 144-foot-long (43-meter) hydrogen-filled balloon was powered by a steam engine that turned a rear propeller. Because the engine had to be light enough for the balloon to fly, it was small—only three horse-power—so the dirigible couldn't fly against a wind. Still, Giffard had shown that controlled lighter-than-air flight was possible.[1]

A dirigible has a large envelope—which contains the lifting gas in one or more gas bags or cells—and a rigid gondola below the envelope that holds the crew, equipment, passengers, and cargo. The engine is mounted in the gondola or elsewhere outside the envelope, and fins on the rear of the envelope keep it stable and allow it to fly straight. If you have ever seen the Goodyear blimp hovering over a sports stadium, you've seen a dirigible.

Early dirigibles used hydrogen as the lifting gas. Hydrogen is the lightest gas there is—much lighter than oxygen—and is easy to generate, but it's also extremely flammable, meaning that the airship's engine had to be placed carefully and the exhaust directed away from the envelope to protect against sparks. Later, the U.S. began using helium, which is

Henri Giffard's airship of 1852.

almost as light as hydrogen and nonflammable, making airships safer.

Although the first dirigible flight took place in France, Germany quickly took the lead in dirigible development. Count Ferdinand von Zeppelin began building large ships in 1898, and the name "Zeppelin" became synonymous with dirigibles.

Zeppelin's dirigibles were rigid airships; that is, the envelope was built around a rigid framework and did not collapse when the gas was removed. His first airship was 420 feet (128 meters) long, and later models grew to twice that size.[2]

Zeppelin's airships began regular passenger service in 1910 and carried more than 37,000 passengers over the next four years. In 1929, the

Graf Zeppelin flew around the world in just over 21 days, and it offered non-stop passenger service across the South Atlantic in 1932–1937. Passengers on this and other large airships enjoyed a level of comfort and service equal to or better than the finest ocean liners.[3] The Empire State Building, completed in 1931, was topped with a dirigible mooring mast to accommodate future passenger flights.

Crew members holding down Zeppelin LZ 127 by its gondola, or passenger compartment.

Alexander Cohrs

Airship designers continued to face the challenge of balancing weight with the buoyancy of the lifting gas. As Zeppelin made larger and larger dirigibles, the necessary trade-offs led to structural weaknesses, and by the late 1930s, it was clear that dirigibles were not safe, practical, or cost-effective for mass transportation. In 1914, a builder could turn out an estimated 34 airplanes for the cost of one Zeppelin.[4] Dirigibles were also difficult to control on the ground, especially in high winds, and they moved slowly in the air.

The dirigible USS *Los Angeles* moored to the U.S. Navy ship *Patoka* in 1931.

U.S. Navy Naval History and Heritage Command

Helium is the second most abundant element in the universe but is found mostly in stars. The United States controlled most of the supply on Earth, where it is a byproduct of natural gas extraction. As a consequence, many countries continued to use hydrogen or a mix of combustible gases in dirigibles, leading to a high risk of explosion and fire. Of 161 dirigibles built over a period of 30 years, 60 were destroyed in accidents due to fire or structural failure.[5] Airplanes, not dirigibles, became the future of air travel.

The *Hindenburg* Disaster

U.S. Navy

This photo, taken during the initial explosion of the *Hindenburg*, shows the 804-foot airship just before subsequent explosions sent it crashing to the ground at Lakehurst Naval Air Station in Lakehurst, New Jersey, on May 6, 1937.

The best remembered dirigible is the *Hindenburg*, manufactured in Germany by Zeppelin in the late 1930s and designed to be "the most luxurious aerial vehicle of the era."[6] Launched in 1936, it offered transatlantic service for 72 passengers with a crew of 40. The *Hindenburg* made 17 round trips that year from Frankfurt, Germany: 10 to New York and seven to Rio de Janeiro, Brazil.[7] A transatlantic trip could take 43 to 78 hours, depending on the route traveled. Not until the early 1940s did airplanes begin making regular transatlantic flights.

On May 6, 1937, the *Hindenburg* caught fire while approaching a mooring mast at the U.S. Naval Station in Lakehurst, New Jersey. The subsequent explosions and crash killed 13 passengers, 23 crew members, and one member of the ground crew.[8] Because this was the *Hindenburg*'s first trip of the new season, many journalists were on hand to record its arrival. Spectacular film footage, photographs, and eyewitness accounts of the disaster raised serious doubts about the safety of these large airships.

Elevators
An Up-and-Down Business

Imagine a city where the tallest buildings are only five stories high. The best apartments, offices, and hotel rooms are on the first floor; the penthouse is the cheapest and least sought-after. The only way to get groceries, luggage, and office supplies to the upper floors is to haul them up the stairs. That's what cities were like before elevators.

The principle of using a hoist to lift heavy objects goes back at least to ancient Greece. In 236 BC, Archimedes, a Greek mathematician, invented a rope-and-pulley tackle to lift objects,[1] and by AD 80, Romans were using a crude rope-and-pulley elevator to lift gladiators and lions to the arena level of the Coliseum.[2] The rope was wound around a drum, and men or donkeys pulled a lever to turn the drum, shortening the rope and lifting the cargo.

By the early 1800s, inventors had added counterweights on the other side of the pulley to help balance the weight of the elevator, and steam or water power was being used to lift materials in mines, factories, and warehouses. But elevators still had one big flaw: if the lift cable broke for any reason, the elevator platform would plummet to the ground.

American inventor Elisha Otis realized that elevators needed a safety brake that would trigger automatically the minute the cable broke to stop the platform from falling, and he set out to develop one. On September 20, 1853, he demonstrated his new invention at the Crystal

The interior of the Brunswick shopping center, Scarborough, North Yorkshire, England, with its glass elevator.

Looking Your Best

Palace Exhibition in New York. Having built a complete elevator in the exhibition hall, he loaded it with freight and then climbed aboard. Once the platform had been raised to its full height, Otis ordered the hoisting rope to be cut with an axe. To the amazement of the crowd, the platform didn't fall. Otis's safety brake worked![3] The new "safety elevator" paved the way for passenger elevators in buildings, and Otis went on to establish an elevator manufacturing company that is still in business today.

The first public passenger safety elevator—installed in a five-story department store on Broadway in New York City in 1857[4]—could carry a maximum load of 992 pounds (450 kg) and was steam-powered.[5] Hydraulic- and electric-powered elevators followed as those technologies became more widespread, enabling elevators to lift heavier loads faster.

Early elevators were operated manually by a trained "driver." It took skill to stop and start an elevator smoothly, without jolting the passengers, and to align it precisely with the exterior floor at each stop.[6] Eventually, elevators were automated, and today it is rare to see a uniformed operator.

Today's elevators feature automated controls and computerized monitoring to ensure the most efficient operation. Automatic scheduling and dispatching reduces the waiting time on any floor during peak

A bank of elevator buttons showing none for the thirteenth floor. In countries where 13 is considered an unlucky number, it's never assigned to a floor in a building.

traffic periods. In the tallest buildings, special express elevators go from the lobby directly to the upper floors without stopping en route. And remote monitoring allows building owners to identify possible problems, such as failing components, before they can interrupt service, ensuring that when someone pushes a button, the elevator will be there.[7]

The Elevator's Impact

Some historians argue that elevators have transformed society as profoundly as the automobile or the electric light.

Along with advances in steel frame construction, elevators led to the development of the modern skyscraper. Buildings without elevators can have no more floors than the number of flights of stairs people are willing to climb. Without elevators, cities would have to spread out horizontally rather than growing vertically. Patrick Carrajat, the founder of New York's Elevator Museum, notes that without elevators we would have one giant city sprawled from Philadelphia to Boston.[8] By building upward, landowners could increase their profits on valuable real estate without having to purchase more land. More people could live and work in the same number of city blocks, and as elevators became more prevalent, building designs changed to accommodate easy access. By the 1930s, architects were putting elevator bays in the centers of buildings,[9] and landlords everywhere began charging higher rents for the more desirable upper floors with their panoramic views and distance from traffic noise.

Elevators also had a significant social impact. People found themselves regularly thrown together with strangers in a small, confined space. Questions of etiquette arose. Should a man remove his hat in the presence of a woman, as he would in a restaurant, or keep it on, as he would if riding a train?[10] Was it necessary to talk to fellow passengers or more polite to ignore them? Psychiatrists began diagnosing a new condition called *claustrophobia*—a fear of confined spaces—the symptoms of which were first noted by elevator passengers.

Subways
Ticket to the Underworld

Clergymen attacked the very idea of subways as satanic. Burrowing underground meant trespassing in the Underworld, the domain of the dead. Other opponents argued that riding through damp, dark underground tunnels would drive people insane, and that digging into the subsoil would release dangerous germs buried with the garbage of past generations.[1] Others said that coal-fired steam locomotives would asphyxiate passengers underground or that tunneling under roads and buildings would cause them to collapse.

Despite all objections, the London Underground, the world's oldest subway, opened in January 1863 and was followed by the first U.S. subway tunnel in Boston in 1897, the Paris Metro in 1900, and the New York City subway in 1904.[2] Cities had one compelling reason for moving people underground: traffic congestion on surface roads.

Paris began enacting ordinances to regulate city traffic as early as the late 1400s. Drivers were forbidden to let their horses trot or gallop, and in 1508, carriages and wagons were restricted to no more than two horses. Pedestrian casualties were frequent.[3] By the mid-nineteenth century, the Industrial Revolution had induced millions of people to flock from the countryside to cities in search of jobs, and growing urban populations led to even more traffic. Carriages, wagons carrying goods for trade, large horse-drawn public omnibuses, carts used to

A modern subway train in Brazil.

Tânia Rêgo/Agência Brasil

collect horse manure and street sweepings, the first automobiles, and pedestrians all contended for the right of way. Traffic engineers in London predicted that the growing number of vehicles would soon make it impossible to get anywhere in the city.[4]

The first subway tunnels were built by a "cut-and-cover" method in which engineers dug up an existing road to create an open trench, shored its sides, roofed it, and rebuilt the road on top. Since cities already owned existing roadways, this approach avoided the cost and public outcry of acquiring land by eminent domain and tearing down buildings.

Subway builders faced a number of engineering challenges. They had to ensure that the road and the houses along it didn't collapse after the tunnel was built. They also had to avoid underground gas, sewer, and water lines, one solution for which was to abandon the cut-and-cover approach for deeper tunnels that went below these obstacles. Much like miners, engineers would dig a vertical shaft, then tunnel horizontally deep underground along the subway route. Over time, inventors developed specialized tunneling machinery.

Despite open vents in their roofs, the air quality in the first subway tunnels was as bad as naysayers had predicted. Accidents occurred because locomotive engineers could not see sig nals through the smoke. But subways were the fastest way to get around a city, and passengers rode them despite the problems.[5]

The development of electric motors in the late 1800s marked a major step forward in subway development, allowing trains to run on

Library of Congress

Digging a New York City subway line between 1910 and 1915. A hand-lettered label identifies the men as "aliens"—i.e., immigrant workers.

electric power transmitted through an overhead line or a third rail on the track. Electrification removed smoke from the tunnels and greatly improved air quality. An American inventor, Frank Sprague, developed a multiple-unit train control system that added a power source to each car. A single motorman could control all the cars from one point, and lighter trains no longer needed a locomotive.[6]

As cities grew, so did their subway systems. Each system became a network of interconnecting lines and stations, and trains became lighter, faster, more comfortable, and more automated. As one example, San Francisco's Bay Area Rapid Transit network, or BART, is fully computerized and automated. The cars are run by electric motors and can reach 80 miles (129 km) per hour while controlled by a central computer system and electric gears on the tracks.[7] Each train carries an operator to handle emergencies.

Today's subways provide a safe, fast, efficient, and affordable way to travel in cities throughout the world.

Subways in Wartime

Imperial War Museum

With German bombers over-head in the fall of 1940, a young woman plays a gramo-phone in an Underground air-raid shelter in north London.

Subways in London, Paris, and other European cities played an important role in World Wars I and II. Many stations were used as air-raid shelters, keeping people safe from German bombs. During the German Blitzkrieg of World War II—the airborne bombing of British cities between September 1940 and May 1941—nearly 200,000 Londoners slept in subway tunnels every night. Parts of the London Underground were also used to store treasures from the British Museum, house military command centers, and serve as factories to manufacture critical parts for tanks and bombers.[9]

Unearthing History

In the course of digging subway tunnels, workers have uncovered important arti-facts of ancient civilizations. Crews have found Roman homes dating back to the second century BC, underground chapels used by early Christians, frescoes, and the foundations of historic buildings. In Moscow, workers uncovered the remains of an underground labyrinth built in the 1560s by Ivan the Terrible. Until then, historians had dismissed tales of the labyrinth as folklore.[8]

Subways in Popular Culture

The apparition of these faces in the crowd;
Petals on a wet, black bough.
—Ezra Pound, "In a Station of the Metro," 1913

Today, people in every major city take the subway as part of their daily routine. Subways have become emblematic of city living, and nowhere more so than in New York City. From Duke Ellington's famous "Take the A Train" to the subway stop on *Sesame Street*, New York subways appear in music, literature, TV shows, and movies depicting city life.

Jennifer Lopez named her debut album *On the 6*, after the number 6 train she rode while growing up in the Bronx. The plot of the 1974 film *The Taking of Pelham One Two Three* (and its 2009 remake) centers on the hijacking of a south-bound number 6 train. TV shows from *I Love Lucy* to *The Odd Couple*, *All in the Family*, and *Seinfeld* have set scenes or whole episodes in subway cars. The expe-rience of being packed together with random people from all walks of life in a (frequently) dirty, noisy metal tube hurtling through a dark tunnel has inspired many poets and writers to use the subway as a motif for individual isolation and the potential danger lurking within our daily routines.[10]

Bicycles
Pedal Power

The invention of the bicycle gave ordinary people an affordable way to travel farther and faster than they could on foot. It also helped women become more independent.

What we know as a bicycle started out as a "walking machine" with a handlebar and two wheels, all made of wood, and no pedals or brakes— similar to the training bikes made for young children today. In 1817, the German Baron Karl von Drais invented this machine to help him get around the royal gardens faster—alternately pushing with his feet and gliding. The British nicknamed this device the "hobby-horse," and it became a popular fad among wealthy young men,[1] but it worked only on well-maintained pathways and quickly wore out riders' boots, limiting its usefulness.

In 1865, a French metalworker connected pedals to the front wheels to create the first real bicycle.[2] Because early bicycles were made entirely of wood or metal, including the wheels, they were often called "bone-shakers," and riding one on a cobblestone street was a truly bone-shaking experience! Early bicycles had a front wheel many times larger than the rear wheel to increase the distance a rider could travel with each turn of the pedals. The rider sat high atop this wheel. Getting on and off was a challenge, and hitting a rut or a stone in the road could mean a quick trip head first over the handlebars.[3]

The development of a fine metal chain and sprocket made it possible to link the pedals to a gear assembly that turned the wheels numerous times for each turn of the pedals. This finally eliminated the need for a supersized front wheel and made the modern bicycle possible.[4] Gear ratios could be adjusted to improve speed and efficiency.

Early rubber tires were solid and not much less bone-shaking than their wooden predecessors. The invention of the pneumatic tire, made of reinforced rubber and filled with compressed air, made biking far more comfortable.[5] By the 1890s, bicycles were widely used by both men and women.

Women in particular benefited from the development of the bicycle. Because it was impossible to ride bicycles in tight corsets, long full skirts, or bustles—all of which were in fashion at the time—bicycling helped push women's fashion toward more comfortable and practical clothes. Bloomers—very wide-legged pants, fastened at the ankle— had been introduced in 1849 and were adopted by the Rational Dress

Roland Fischer

Antique wooden "walking machines" on display in Zurich, Switzerland, 2014.

Mᵐᵉ Jeanne LABATOUX

A bloomer-clad bicycle rider.

Society, founded in 1881. Originally considered shocking, bloomers were accepted as suitable attire for bicycling by 1895.[6] More importantly, bicycles gave women the freedom to travel about on their own without needing a carriage and driver.[7]

Since 1990, rising gas prices and concern for the environment have caused a renewed interest in bicycling. Today, many cities provide special traffic lanes and bike paths for cyclists so that people commuting to work can ride safely. And cities throughout the world[8] have instituted bike-share programs that make bicycles available in specified locations

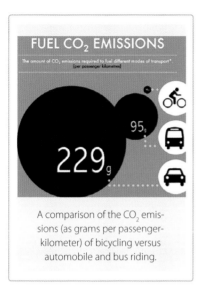

FUEL CO$_2$ EMISSIONS

The amount of CO$_2$ emissions required to fuel different modes of transport*.
(per passenger kilometer)

95$_g$

229$_g$

A comparison of the CO$_2$ emissions (as grams per passenger-kilometer) of bicycling versus automobile and bus riding.

for short-term use. Riders pay a small fee to use a bike, then leave it at any convenient bike-share location when they finish their ride.

Today's bicycles are far lighter, stronger, faster, and more comfortable than ever before, thanks to such features as gelled seats, ergonomic fitting, and sophisticated shock absorbers. And they still provide transportation, fun, exercise, and adventure for riders all around the world.

Which Bike for You?

Anthony DeLorenzo

Riding a "fat bike" in the snow.

Drop by a bicycle store and you'll be amazed to see the range of options available. Bicycles come in a wide variety of models tailored for specific uses and environments. Do you plan to ride at an easy pace on flat roads? A comfort bike with wide tires and a well-cushioned seat will do the trick, or perhaps you'll choose a recumbent bike that allows you to lean back, supporting your back and relieving weight and pressure from your buttocks—and you can add a windshield or even a canopy for shelter and aerodynamic efficiency. If you're looking for speed, you'll want a racing bike, which has a lightweight frame, drop handlebars for an aerodynamic riding position, narrow, high-pressure tires to decrease road resistance, and multiple gears. BMX bikes are designed for doing stunts and tricks and for racing on dirt tracks. Mountain bikes or all-terrain bikes are built to handle rough, uneven ground. Fat-treaded tires and sturdy frames help absorb the shock of a rough ride, and some bikes come with suspension systems. Low gearing helps a rider stay upright at very slow speeds. "Fat bikes," built for riding on snow, have tires that are nearly twice as wide as mountain bikes, with extra-large treads to provide traction. There are even specialized bikes for those who want to ride on water.

Bicycle Racing

Bicycle racing became popular in the late 1800s in Europe and the United States. Races were often held on specially built wooden tracks, could last up to several days, and drew large crowds. The first Olympic Games in 1896 included bicycle racing as a sport, and the first Tour de France, now a world-famous annual bicycle race, was held in 1903. The race was

Competitors in the Tour de France of 1906.

Bibliothèque Nationale de France

sponsored by *L'Auto*, a French newspaper printed on yellow paper; hence the yellow jersey that is still worn today by the lead rider in each leg of the race.[9]

Multimodal Transportation

Bicycling is great for short distances, but few people—even those committed to decreasing their carbon footprint—have the time or energy to commute 20 miles to work on a bike. Luckily, cities have recognized the need to integrate bicycles into their transportation networks and are taking steps to do so. Many city buses are equipped with bike racks. Subway and light rail systems are expanding the hours during which riders may bring bicycles aboard and offering bike racks or lockers at stations. City transportation planners are working to connect bike paths and lanes with bus and subway stops to create seamless end-to-end routes. Bicycle manufacturers that offer folding bikes are doing their part, too.

Motorcycles
Bicycles on Steroids

Many early two- and three-wheeled vehicles were built as test platforms for new automobile engines, but the motorcycle, a kind of hybrid of bicycle and car, evolved as a vehicle of its own. Many early motorcycle designers and builders worked on automobile development as well.

An American named Sylvester Roper invented what can be considered the first motorcycle in 1867. His two-wheeled vehicle was powered by a steam engine that burned coal. The first motorcycle with a gasoline engine was built by Gottlieb Daimler in 1885.[1] Daimler mounted his engine on a wooden bicycle frame with the pedals removed,[2] and he included a "spring-loaded outrigger wheel" on each side to add stability.[3] In other words, the first gas-powered motorcycle had training wheels!

Daimler's motorcycle had wooden wheels with iron rims, but in 1892, Alex Millet designed a better version with pneumatic tires and an improved engine built into the rear wheel. In 1895, De Dion-Bouton, a French automobile manufacturer, developed a small, lightweight, four-stroke engine that became the model for motorcycle manufacturers around the world.[4]

An early problem for motorcycle builders was where to place the engine. This was an important question, affecting the motorcycle's center of gravity as well as the rider's comfort. Pretty much every position was tried: built into the rear wheel, towed behind the bike, clipped below the seat, or mounted over the rear or front wheel.[5] Eventually, it became clear that the best place for the engine was as close as possible to the bottom bracket on the frame. This kept the center of gravity low, improving the bike's stability, and kept the engine away from the rider's knees.[6]

By the early 1900s, companies had built factories and were producing motorcycles for sale to the public. Motorcycles were less expensive to buy, used less gasoline, and were more reliable than the automobiles

Ewald Gabard

Modern Harleys pass an antique bike at Rock the Roof 2013 in Schladming, Styria, Austria.

of the day.[7] In the U.S., the Hendee Manufacturing Company produced the first Indian motorcycle in 1901, and Harley-Davidson introduced its first motorcycle in 1903.[8] Both incorporated the De Dion-Bouton engine developed in France.

Early motorcycles had only one gear and required pedals for manual power when starting the engine or climbing hills. To start, the rider had to either pedal very fast to jump-start the engine or run alongside the bike to get it up to speed and then jump on. If pedaling failed to give the engine enough extra power to climb a hill, the rider would have to get off and push. Motorcyclists planned their routes to avoid steep hills.[9]

In 1906, the Yonkers, New York, police department became one of the first to replace bicycles with motorcycles. Motorcycles could keep up with or outpace horse-drawn carriages and were equipped with speedometers, making them a superior tool for enforcing speed limits. In World Wars I and II, soldiers on motorcycles conducted reconnaissance, delivered messages to front lines, and carried wounded comrades to medical stations. In World War II, folding motorcycles were parachuted to agents behind enemy lines.[10]

A 2.5-horsepower British Matchless from 1905 shows its bicycle origins. Like many other motorcycle manufacturers, Matchless began in the bicycle trade before producing its first motorcycle in 1899. It went on to become one of the most famous names in the British motorcycle industry.

World War II also brought improvements in airplane engines, which had to be light and efficient, and motorcycle manufacturers incorporated these improvements after the war. GIs returning from duty in Japan introduced the first Honda motorcycles to the U.S.,[11] and in 1946, an Italian manufacturer introduced the Vespa, a lightweight motor scooter.[12] Whether you were looking for speed and excitement or cheap, reliable transportation, there was a bike for you.

Today, people in most developed countries use motorcycles primarily for recreation. Organizations like the American Motorcyclists Association and the Federation of European Motorcyclists Associations link motorcycle groups and promote safe riding and a positive public image for motorcyclists. In developing countries, motorcycles and motor scooters continue to be a primary means of transportation.

A replica of Daimler's 1885 motorcycle at the Mercedes-Benz Museum in Stuttgart, Germany.

Wladyslaw

The Biker Image

By the 1960s, motorcycles in the U.S.—especially the Harley-Davidson brand—had become associated with bad behavior. A 1947 gathering in Hollister, California, with approximately 4,500 attendees, erupted into a drunken melee. Bikers raced through the streets and fought with police. A photo of the uproar made the cover of *Life Magazine*. In 1953, *The Wild One*, a movie starring Marlon Brando, romanticized the image of the outlaw biker outfitted in a leather vest, tattoos, and chains. A group called the Hell's Angels became the best-known motorcycle club. Even though the "outlaws" have always been a small percentage of motorcyclists, the image persists today.

Rickshaws
Human-Powered Vehicles

Rickshaw pullers and their passengers in Kolkata (Calcutta), India, were up in arms! The year was 2005, and the Kolkata Municipal Council had just voted to stop reissuing licenses to hand-pulled rickshaw drivers. Chief minister Buddhadeb Bhattacharjee wanted to "liberate" the human beasts of burden and abolish this "inhuman practice dating back to colonial times."[1]

The word *rickshaw* comes from the Japanese word *jinrikisha*, which means "human-powered vehicle."[2] A rickshaw is a chair-like structure mounted on two large wheels, with two long poles attached to the front of the chair. The rickshaw puller, or runner, grasps the poles and pulls the vehicle while walking or running through the streets.

The first rickshaw was built in 1869 by a European missionary in Yokohama, Japan, to carry his invalid wife through city streets.[3] The Japanese had few horses, but human power was plentiful, and rickshaws quickly became a popular form of transportation. By 1872, there were an estimated 40,000 rickshaws licensed in Tokyo alone.[4]

Japanese emigrants exported the rickshaw to China around 1874, and from there it spread throughout Asia, India, and South Africa. Rickshaws became a favored means of transportation among the wealthy and officials of colonial governments in many countries.

Arne Hückelheim

A human-powered rickshaw in Kolkata, India.

The life of a rickshaw puller was not easy. According to Margaret Simpson:

> Often the vehicle he pulled was his whole world where he ate, slept and worked. His meagre possessions were kept in a compartment under the seat. These might have included a spare pair of straw sandals, a pipe and tobacco pouch and a paper lantern which he lit and hung on the rickshaw shafts at night. Rickshaws always travelled in single file, and the runner in front called out the particulars of hazards to his comrades coming behind.... He ran at an easy gait and if the person being drawn was overweight or the route hilly, a second runner joined him either in pulling or pushing the rickshaw and the passenger was requested to pay an extra amount.[5]

By 1938, more and more people were traveling by car, bus, taxi, or train, and only 13,000 rickshaws were licensed in Japan.[6] In 1949, Mao Zedong, chair of the Chinese Communist party, banned the rickshaw as a symbol of working-class oppression by the rich and privileged. India and many other Asian countries followed suit.[7] In recent years, human rights groups have supported efforts to abolish rickshaws in the few places where they are still used, arguing that they exploit poor workers. One official was quoted as saying that it is offensive to see "one man sweating and straining to pull another man."[8] That sentiment led to the Kolkata ban.

Proponents of the Kolkata rickshaws argued that hand-pulled rickshaws provide low-cost transportation for people who cannot afford other ways to travel. They also provide jobs for large numbers of refugees and recent arrivals from the country's villages. Many people were concerned that banning rickshaws would leave large numbers of the poor without jobs. Travel by rickshaw is often faster than by car in congested streets, and rickshaws can traverse narrow alleyways and streets

A rickshaw rider with parasol between 1880 and 1900. Walking ahead of the rickshaw puller is the man who will relieve him.

Library of Congress

that aren't accessible to other vehicles. In the monsoon season, with its heavy rains, rickshaws can move through flooded streets where the water is too deep for an automobile. Some families contract with a regular rickshaw puller to take their children to and from school each day. In December 2011, the Kolkata Municipal Council reversed the ban and resumed issuing hand-pulled rickshaw licenses.[9]

Today, human-powered rickshaws can be seen only in a few places, usually as tourist attractions. Kolkata remains the only city where rickshaws are still in use for daily transportation. The estimated 6,000 licensed *rickshaw wallahs*, as the pullers are called, continue to resist further attempts to ban their vehicles.[10]

Rickshaw Pullers on Parade

A modern Durban, South Africa, rickshaw puller in costume.

Rickshaw pullers along the Marine Parade in Durban, South Africa, often make more money posing for tourist photographs than by pulling their vehicles through the streets.

In the early 1900s, as rickshaws became more popular in Durban, the government required pullers to wear a uniform. The plain government-issued fabric trimmed with a single band of red braid was far too dull for the majority of pullers, who were from the Zulu tribe. They began to add extra braid and other decorations and put horns on the hat to show that they were "strong as an ox." Competition soon developed to design the most colorful costume, resulting in extravagant beaded headdresses and body coverings. Today, it's a wonder that some of these pullers can find the strength to carry their heavy costumes, much less pull a rickshaw.[11]

Cruise Ships
Floating Cities

Cruise ships may travel to exotic locations, but for some people the ship itself is the destination, and why not? Passengers can enjoy a variety of restaurants, see a Broadway show, shop, or work out. They can swim in a pool or zoom down a waterslide, gamble in a casino, go to the movies, or enjoy a luxury spa treatment without leaving the ship. Some ships even offer ice skating rinks, rock climbing walls, tennis courts, and miniature golf courses.[1] Cruise ship companies are always adding new activities that will convince people to sign on for a vacation trip.

Cruise ships differ from ocean liners (Chapter 28) and from most other vehicles in that they don't move passengers from point A to point B. Rather, a cruise ship visits one or more destinations and then brings its passengers back to the port of embarkation. Some cruise lines even advertise a "cruise to nowhere," in which the ship goes out to sea for two or three days and then returns to port, the only purpose being to relax, enjoy on-ship activities, and sometimes gamble onboard in international waters.

Modern cruising began in 1844 when the Peninsular & Oriental Steam Navigation Company (P&O) began advertising sea tours from Southampton, England, to ports along its cargo routes, such as Gibraltar, Malta, and Athens. Later the company expanded its offerings to include cruises to Alexandria and Constantinople. As more people signed on,

P&O commissioned larger and more luxurious ships to accommodate them. The first vessel designed exclusively for cruising, the German ship *Prinzessin Victoria Luise*, was commissioned in 1900.

In 2010, the North American cruise industry included 205 ships that carried nearly 15 million passengers, of which almost 10 million sailed from U.S. ports. The most popular destination was the Caribbean, followed by the Mediterranean, Europe, and Alaska. Cruises can last three days to several months for around-the-world trips; the average is seven days.[2] Ocean liners may have disappeared from the seas, but the major cruise lines commissioned an average of ten new ships per year between 2001 and 2009. One of the largest, Royal Caribbean's *Freedom of the Seas*, is 1,119 feet (341 meters) long and exceeds the gross tonnage of the last remaining ocean liner, *Queen Mary 2*.

Quinn Norton

A waterslide and swimming pool on a cruise ship.

The Windjammer Barefoot Cruise Lines tall ship *Fantome* off Dominica. The *Fantome's* sails were more for show than propulsion, most of which was provided by her diesel engines. The 282-foot steel-hulled vessel proved no match for Hurricane Mitch in October 1998. As Mitch spun up out of the southern Caribbean, *Fantome* disembarked her passengers and put to sea with a skeleton crew to dodge the storm, but Mitch swerved to meet the ship with 180-mile-per-hour winds and 50-foot seas. Ship and crew vanished without a trace.

An average cruise ship carries about 2,200 passengers—the largest can hold more than 5,000—with approximately one crew member for every two passengers.[3] Many ships resemble small floating cities. Not everyone wants to travel with such a huge crowd, however, so some cruise lines specialize in smaller ships. Two popular options are river cruises and tall ships, often called *windjammers*.

Because many cities and historic sites grew up around major rivers, river cruises can be an excellent way to tour. Routes include the Rhine, Nile, Yangtze, Mekong, and Mississippi Rivers. Popular cruise lines such as Viking River Cruises carry between 60 and 200 passengers on each vessel and provide walking tours of important sites at every stop.[4]

Tall ships are multimasted sailing vessels that carry up to 200 passengers and offer a casual environment with fewer luxuries. Passengers are often invited to help with shipboard duties like raising the sails. Tall ships can access smaller ports that large cruise ships can't enter.

Cruise ships face the same dangers as any vessel at sea, including fires, unintentional groundings, and collisions. In January 2012, the *Costa Concordia* hit a rock off the coast of Italy and capsized several hours later with the loss of 32 passengers. Another cruise ship was attacked by pirates off the coast of Somalia in 2009,[5] and more than one passenger has been lost overboard. But the more common risks are like those of any vacation travel, including petty theft, assault, and gastrointestinal infections such as norovirus. Cruise lines provide onboard security, cleaning services, and medical care to limit such risks and ensure that passengers enjoy the relaxing vacation they expect.

Educational and Theme Cruises

Un-Cruise Adventures

Passengers from the adventure cruise ship *Safari Explorer* kayak around floating ice in Alaska.

Many cruise lines offer special educational programs for passengers who want more from a trip than simple relaxation. Combine a cruise of the Greek islands with expert lectures on ancient Greek culture and guided tours of historical sites. Spend your week in the Mediterranean sampling the region's wine and food. Or learn about Viking history while cruising the waters the Norse raiders once sailed. Some cruise lines offer a few classes during the course of the trip. Others—such as Smithsonian Journeys, which partners with the Smithsonian Institutions, and Road Scholar, which caters to older adults—make education the primary focus of the trip.

Theme cruises are organized around a shared interest. Whether it's chess, NASCAR racing, the novels of Jane Austen, yoga and holistic health, tattoos, or a shared love of Monty Python, a theme cruise lets passengers enjoy learning more about their favorite topic and meeting others who share their interest.

Airplanes
The Sky's the Limit

By the mid-1800s, humans had achieved lighter-than-air flight with hot air balloons (Chapter 23) and dirigibles (Chapter 29). Balloons were at the mercy of the winds; dirigibles were self-propelled and steerable but slow, unwieldy, and dangerous.[1] A stronger, more reliable vehicle was needed to move passengers and cargo by air, and that meant heavier-than-air flight.

Inventors had built unpowered gliders that could fly short distances when launched from a hilltop, but no one had kept a plane in the air for long. And successful flight required more than keeping a plane in the air. The pilot needed a way to control the plane's altitude, speed, and direction.

Wilbur and Orville Wright, two American brothers, had observed birds in flight and built bicycles, and they understood the need for flexibility. The Wrights designed a plane that a pilot could control by bending the edge of one wingtip up and the other down (later called "wing warping"), thus providing more lift on one side of the aircraft than the other. Wing warping allowed the pilot to balance the plane or induce it to *bank*, or turn, as needed. The Wrights also added vertical tail surfaces and a rudder that worked with the wing warping to keep the aircraft steady.[2] Their first successful flight on December 17, 1903, at Kitty Hawk, North Carolina, lasted only 12 seconds, but later flights that same day stayed airborne almost a minute.

The Wright brothers' first successful flight, December 17, 1903.

Library of Congress

The Wright brothers continued to improve their designs, and by 1905, they could fly 25 miles in just under 39 minutes. Inventors in France and England were meanwhile working on their own designs. Early designers learned from each other's successes and failures while competing to set new records for flight times and distances.

Early planes were made of wood, usually ash and spruce, with wire bracing and a cover of silk or linen. Engines were underpowered, and engine failures were common. Pilots flew in an open cockpit and could not fly at night or in bad weather. Longer flights were frequently interrupted to refuel or make critical repairs.

Calbraith Perry Rodgers made the first successful flight across the U.S. in 1911. He averaged almost a mile a minute in the air, but the trip from New York to California took 89 days. By the end of the flight, Rodgers "had endured five major crashes, numerous lesser mishaps on takeoff or landing, engine failures in midair and a hospital stay. Of his original plane, only the rudder and the oil drip pan remained; everything else had been replaced at least once."[3] Rodgers flew the last leg of the trip with a broken ankle, his crutches strapped to the wing of the plane.

By the early 1940s, planes could carry 30 or more passengers across the Atlantic or Pacific with baggage and additional cargo, but air travel was still expensive and difficult. Passengers flew in unheated cabins and were bombarded by engine noise and vibration. Heavy turbulence made almost everyone airsick.

Air travel has come a long way since the Wright Brothers first took flight.

peter s c/o Shutterstock.com

By the 1950s, planes had become faster, safer, and more comfortable, in large part due to the introduction of jet engines. Planes were more fuel-efficient and carried more passengers. Ticket prices fell, and more people began to fly. In 1956, more people crossed the Atlantic by plane than by ship.

Air travel has made the world a smaller place. The benefits are many, but not without cost. Exotic viruses and invasive species of plants and insects can travel more easily via airplane to afflict new regions, and some critics say that by spreading dominant cultures so widely, we are erasing important and interesting differences around the world.

How High Did They Fly?

Library of Congress

Children on a beach watch Louis Blériot's plane as he flies over the English Channel on August 3, 1909.

Early flights got off the ground, but not by much. In 1909, Henri Farman flew more than 100 miles (180 km) at a height of 12 feet (4 meters) around a circular track,[4] and Louis Blériot made his historic first flight across the English Channel at an altitude of 250 feet (76 meters).[5] In comparison, the typical cruising altitude of today's airliners is 30,000 to 40,000 feet (5.7 to 7.6 miles or 9.1 to 12.2 km). Flying at such altitudes puts a plane above most weather and turbulence, and the thinner air means less wind resistance and better fuel efficiency. It also allows more vertical separation between planes, reducing the risk of midair collisions.[6]

History of Flight Attendants

The first-class cabin on a Scandinavian Airlines DC-8 in 1969.

Ships and passenger trains had long employed stewards to serve food, maintain cabins, and assist passengers. When dirigibles began flying, they provided stewards as well, and in the 1920s, the United Kingdom's Imperial Airways had them on board.

Boeing Air Transport introduced the concept of female flight attendants in 1930. The airline hired eight trained nurses to assist the large number of airsick passengers. These nurses, called stewardesses, were supposed to make passengers feel safe and comfortable. Other airlines followed, and by the 1950s, the focus had changed, with stewardesses being hired for youthful good looks rather than motherly demeanors.[7] Beautiful young women in uniform were part of the airlines' attempt to make flying a glamorous endeavor and appeal to business travelers, the majority of whom were men. Applicants were required to meet age, height, and weight restrictions and to be single and female. By the 1990s, all such restrictions had been lifted, and today's flight attendants—male and female—are a diverse, well-trained group of professionals dedicated to passenger safety and comfort.

Airplanes at War

At the start of World War I in 1914, airplanes were still in early development. The military didn't see much use for them beyond tracking and photographing enemy troop movements and locating targets for attack. It wasn't long, however, before planes took a much more active role.

Generals began looking for ways to shoot down enemy planes to stop their surveillance. At first pilots, traveling in open cockpits, shot at each other with pistols. Development of stronger, more maneuverable airframes and more powerful engines led to aircraft being outfitted with machine guns.[8] There was only one problem: a machine gun mounted in front of the pilot would shoot off the propeller blades. Anthony Fokker designed a series of pistons that kept the gun from firing when the propeller blades were directly in front of it, creating the first true fighter plane.[9]

Bombers followed a similar evolution. Early planes were light with small engines, and bombs were heavy; a pilot could carry only one. The development of larger, multi-engine planes increased the number of bombs in each plane, and better targeting tools improved the bombers' accuracy.[10]

World War II brought more advances in airplane size, speed, and maneuverability, including introduction of the jet engine and the use of radar, which used radio waves to track the positions of incoming aircraft. For the first time, people on the ground could have advance warning of incoming enemy air raids.

Innovations from both wars were incorporated into commercial aircraft, and military-trained pilots were valuable assets to airlines seeking to expand.

Snowmobiles
Crossing Winter Landscapes

A Lombard log hauler in the Maine woods.

The first self-propelled vehicle designed to travel over snow wasn't built until five years after the Wright brothers' first flights. The motorized sled that Alvin Lombard built in 1908 weighed nearly 20 tons[1] and looked more like a steam locomotive than a modern snowmobile. Designed to haul logs out of Maine forests in winter, it had skis in front and a continuous rubber track—similar to those used on a tank—on the back.[2]

Smaller snow vehicles followed. Early models included a Model T Ford converted to run on skis and a track. The inventor, an American named Virgil White, coined the word *snowmobile* to describe his new vehicle. Joseph-Armand Bombardier, a Canadian, was only fifteen when he designed a sleigh with a Model T engine in 1922.[3] A wooden airplane propeller mounted on the engine's drive shaft propelled it.[4] And in 1927, Carl Eliason of Wisconsin received a patent for his motor toboggan. Eliason's vehicle was "a wooden toboggan fitted with two skis, which were steered with ropes, powered by a 2.5-horsepower Johnson outboard motor, and was pushed by an endless steel cleated track."[5]

Joseph-Armand Bombardier continued to create and patent new designs. In 1942, he opened l'Auto-Neige Bombardier Limitée in Valcourt, Québec, to build tracked vehicles designed for traveling over snow, and in 1959, he began building and selling the Ski-Doo, considered the first modern snowmobile. The Ski-Doo was lighter and easier to maneuver than earlier versions and quickly became popular.[6] Today, his company, named BRP, builds snowmobiles as well as personal watercraft and all-terrain vehicles for recreational use.[7]

As snowmobiles became more popular, other companies such as Polaris and Yamaha began building and selling them. Snowmobiling

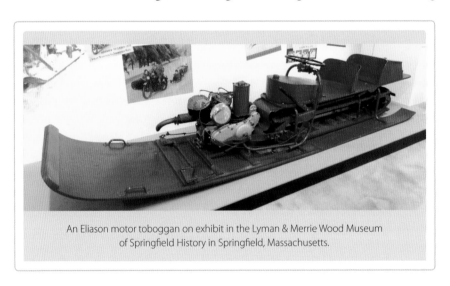

An Eliason motor toboggan on exhibit in the Lyman & Merrie Wood Museum of Springfield History in Springfield, Massachusetts.

became a major recreational activity in Canada, the northern United States, northern Europe, and Russia, and there are an estimated four million snowmobile riders worldwide. Parks and recreational areas provide groomed trails, and day trips are popular. More than 3,000

Stanco

The Adam Renheim Swedish Championships Snocross, 2014.

snowmobile clubs worldwide provide opportunities for snowmobile owners to gather, organize rides, and share information. Many members help maintain the trails they use.

Snowmobiles are often the primary means of winter transportation in remote arctic and subarctic regions, and they are widely used by search-and-rescue teams, emergency responders, public utility workers, environmental and wildlife scientists, ranchers, surveyors, and ski patrols. Some cross-country ski areas use snowmobiles to groom trails, and cross-country ski and dogsled race officials also use them.[8]

Rider safety is a concern, just as it is for motorcyclists and ATV operators. Government agencies at the federal and state or province level have developed regulations for snowmobile registration and safety, and clubs and other organizations provide safety training courses for new riders. When ridden safely, snowmobiles provide both a reliable way to travel and a wonderful opportunity to explore winter landscapes. But with snowmobiles, cross-country skiers, dogsleds, and winter hikers all using the same trails, conflicts do occur.[9] Environmentalists and skiers complain that snowmobile riders damage sensitive habitats and increase noise pollution.

Environmental Impacts of Snowmobile Use

Studies have found that snowmobiles "produce significant impacts on animals, plants, soils, air and water quality, and the ecology of entire winter ecosystems." Heavy snowmobile traffic can separate wildlife from their preferred grazing areas.[10] Snowmobiles also emit significantly more hydrocarbons, nitrogen oxides, and carbon dioxide than automobiles. One study reported that while 16 times as many cars as snowmobiles enter Yellowstone National Park each year, snowmobiles contribute 90 percent of the total hydrocarbon emissions.[11]

Snowmobile emissions also alter the chemistry of the snowpack. Pollutants are absorbed from the air, stored in snow, and released in a concentrated burst during the spring melt, creating a significant threat to surface water and groundwater.[12] In arctic regions, snowmobiles "compact fragile tundra and permafrost ecosystems, cause permafrost to melt prematurely and generally increase soil temperatures."[13]

Snowmobile manufacturers dispute many such findings, contending that the arguments against snowmobiles are being made by cross-country skiers who don't want to share the wilderness.[14] But manufacturers have also taken steps to reduce emissions and noise pollution in newer snowmobile models.

Recreational Vehicles (RVs)
Cruising on Land

Camping can be a fine way to enjoy a vacation: inexpensive, close to nature, flexible, open to spontaneity. But often the thrill of "roughing it" evaporates after a day or two without a refrigerator, a microwave, or a good mattress. For those who want to take their creature comforts with them, recreational vehicles (RVs) provide an enticing solution.

In North America, the term RV refers to a motor vehicle or trailer that contains most of the amenities found in a home. In other countries, these vehicles may be called caravans, camper vans, or motor homes. RVs provide a comfortable, movable living space that can be used as a vacation home, a permanent residence, or a mobile office. RV touring is a lot like coastal cruising on a boat. In both instances, the amenities go along for the ride. The kitchen is called the galley. If the RV tows a car, that car is called a dinghy. When the RV is hooked up to electrical and sewage services at a commercial campground, it's said to be docked.

Wagons to live in, rather than to carry goods, developed in France about 1810. Showmen traveling between country fairs began using similar wagons in England in the 1820s, and by the 1850s, the European Romani (impolitely known as gypsies) had begun living in them.[1]

The earliest RVs were customized automobiles and pickup trucks. Cars in the early 1900s had large running boards—steps fitted under the side doors to help people climb in and out—where a box that opened into a makeshift kitchen or folding cot could be carried. Rear seats could fold out to become beds, and a canvas awning attached to the side of the car provided additional shelter. Camping equipment could be carried in a small wooden trailer towed behind the car.

By the late 1920s, several manufacturers began building camper bodies that could be mounted on a Model T or other auto chassis to create a "house car," and travel trailers designed as moving residences from the wheels up appeared in the 1930s. In 1936, the Airstream Trailer Company introduced the aluminum "Clipper," which could sleep four

German Federal Archive

A German family "house truck" from 1922.

people, carried its own water supply, and included electric lights and an air conditioning system that used dry ice. Larger and more luxurious vehicles followed.

A 1958 Airstream in the RV/MH Hall of Fame in Elkhart, Indiana.

At the same time, miniature travel trailers called teardrops were also becoming popular. The teardrop—named for its shape—was approximately four feet wide by eight feet long and four feet high, small and light enough to be towed by a passenger car or motorcycle. The average teardrop provided room for two people to sleep, and the sleeping cabin could be used to store cargo while traveling. A hatch in the back could be raised to access a kitchen that contained an ice chest, portable stove, and cooking pots and utensils. *Popular Mechanics* magazine published plans for building teardrops, and many owners built their own.

During World War II, essential materials were diverted to the war effort, and gasoline rationing limited travel. Manufacturers of travel trailers suspended production and turned instead to turning out stationary house trailers for war workers. When production of travel trailers resumed after the war, they looked like motorized versions of house trailers, and the result was the modern RV or motor home.[2]

Today, recreational vehicles range from small, easily towed pop-up campers to larger fifth-wheel trailers designed to be hooked to the bed

The kitchen, or galley, in a modern RV.

of a pickup truck, and to bus-size vehicles with all the amenities of a modern home, including wood cabinets, granite countertops, leather sofas, flat-screen TVs, washer/dryer combinations, and walk-in showers. Some large models even include a garage for a Smart Car so that travelers can have a car at their disposal while the RV is parked at a campsite.[3]

Helicopters
Da Vinci's Dream

Helicopters can take off and land vertically, hover in one spot, fly backwards or sideways, and remain aloft at much slower airspeeds than a typical airplane—all of which means that they can be used in congested areas and for tasks a plane can't do. They deliver supplies to remote outposts, oil rigs at sea, and refugee camps; they rescue stranded mountaineers and mariners; and they fight wildfires. They are used to monitor and report on city traffic and to deliver accident victims to trauma centers. It's hard to believe that even as recently as the 1940s, many people questioned their practical value.

The first model of a helicopter was a Chinese toy in use as early as 400 BC[1]—a lightweight stick topped with feathers or bamboo blades. Hold the stick upright between your two palms, spin it fast to create lift, let it go, and watch it rise into the air on its own. Versions of this toy spread through Asia and Europe and, two millennia later, helped develop the Wright brothers' interest in flight.[2]

Achieving vertical flight in an aircraft was much harder than that, however—even harder than building a successful airplane. Airplanes would be crossing the oceans before someone developed a functional helicopter.

Leonardo da Vinci, who painted the *Mona Lisa* and *The Last Supper*, was also an inventor, and in the late 1400s, he designed several helicopters, including a "Helical Air Screw" powered by four men turning cranks. Da Vinci never built the air screw, and modern scientists believe it would have been too heavy to fly.[3] In 2013, however, a team using one of da Vinci's other helicopter designs and ultra-lightweight materials achieved human-powered flight.

Science fiction author Jules Verne's 1886 novel *Robur the Conqueror*, also known as *The Clipper of the Clouds*, features a huge flying vessel called the *Albatross*, with 37 masts, each with an air screw at the top for vertical flight, and propellers on the bow and stern to move it backward and forward. Inventors of that time were turning out an amazing variety of equally grand and complex helicopter designs and small working models, but none flew successfully. As with early airplanes, pilots could not control the motion of the few designs that did manage to get airborne.

Luigi692

A CH7 Kompress C helicopter awaiting delivery in Asti, Italy.

It wasn't until the late 1930s that Anton Flettner, a German aeronautical scientist, developed the first true helicopter.[4]

Air streaming over the curved upper surface of a helicopter's spinning rotor blades travels farther and faster than the air flowing underneath, and that reduces the air pressure above, creating lift. (Air flowing over and under an airplane's wing generates lift in this way, too.) The faster the blades rotate, the more lift they create, and the blades rotate fast even when the helicopter is moving slowly, allowing it to take off vertically and hover.

Just as with so many other motorized vehicles, the helicopter's practicality depended on an engine that was light but powerful. The internal combustion engine marked a big step forward, but piston-driven combustion engines were slow and caused vibration. The turboshaft jet engine, created in the 1950s, finally provided the solution, producing the same horsepower at half the weight of a piston engine, with a much smoother ride.[5]

The pilot and copilot at the controls of a British Royal Navy Seaking Mk4 helicopter on operations over Helmand Province, Afghanistan.

LA(Phot) David Bunting/MOD

The second, more difficult design challenge was controlling the movement of the aircraft in the air. On a helicopter with a single rotor, the torque of the spinning rotor causes the body of the helicopter to spin in the opposite direction. Inventors needed an anti-torque control to enable the aircraft to fly level and straight. After much trial and error, engineers found the right combination of a horizontal rotor on the helicopter's tail, hinged rotor blades, and pilot controls to solve this problem.

Proving da Vinci's Designs

Leonardo da Vinci created several designs for a helicopter, all human-powered. Most modern experts agreed that none of the designs would be able to achieve enough lift to fly—but they were wrong.

In 1980, the Sikorsky Aircraft Corporation announced the AHS Igor I. Sikorsky Human Powered Helicopter Competition. The first team to develop a human-powered vehicle that could fly for at least 60 seconds, rise to an altitude of at least 3 meters (about 10 feet), and remain within a horizontal area no larger than 33 feet (10 meters) square would win a $250,000 prize. The winner had to be certified by the Fédération d'Aviation, the governing body of international aeronautical prizes, before the prize could be awarded.[6] On June 11, 2013, Aerovelo, a Canadian startup company founded by Todd Reichert and Cameron Robertson, won the prize. Their craft, named *Atlas*, flew in an empty soccer stadium for 64 seconds and achieved a maximum altitude of 3.3 meters.[7]

(Continued on next page)

(Continued from previous page)

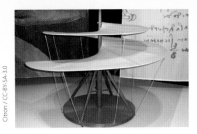

A model of Leonardo da Vinci's aerial screw. Da Vinci believed that four people pushing the bars at the lower level would rotate the wing and cause the craft to lift off the ground.

Built by hand from carbon fiber, *Atlas* weighed 120 pounds (54 kg) and had a wingspan of nearly 154 feet (47 meters). The pilot flew by pumping a set of bicycle pedals to wind spools of string that turned four rotors.[8]

Aerovelo proved that at least one of da Vinci's concepts was workable given the availability of sufficiently light, strong materials.

Plucking a Man from a Crane

A search-and-rescue swimmer is lowered from an SH-60F Sea Hawk helicopter flying near the Nimitz-class aircraft carrier USS *Carl Vinson* during a demonstration in the Atlantic Ocean, 2010.

Daring helicopter rescues make for exciting news reports and high-adrenaline scenes in adventure movies. Helicopters pluck people off sinking ships, burning buildings, and high mountaintops, often with only seconds to spare. But a rescue in Kingston, Ontario, Canada, in December 2013 took heart-stopping action to a new extreme.

A fire at an apartment complex under construction was burning out of control and had spread to the roofs of adjoining buildings. All the residents had been safely evacuated, but an enormous construction crane was stranded in the middle of the fire, and the crane operator had crawled out to the end of the boom to escape the flames. If not rescued quickly, he would burn to death.

Sergeant Cory Cisyk and his team, military rescue technicians for the 424 Transport and Rescue Squadron, carried out a daring maneuver in thick smoke to save the crane operator.[9] Dangling from a cable hanging from the team's helicopter, Cisyk was able to loop a belt around the man's body. The helicopter crew then hoisted the two of them to safety.

Pedicabs

A Cycle-Rickshaw by Any
Other Name Would Ride as Sweet

As bicycles became more widely used, someone got the idea of hooking one to a rickshaw. After all, wouldn't it be easier and faster to pedal than to pull a rickshaw on foot? It's not clear exactly when the first cycle-rickshaw was built, but by the 1940s, cycle-rickshaws had replaced pulled rickshaws in most of India and Asia.[1]

The first cycle-rickshaws were two-wheeled bicycles connected to the standard rickshaw carriage. Soon, however, the back wheel of the bicycle disappeared, leaving a three-wheeled vehicle. In most cases the passengers sat in the carriage behind the driver, although some cycle-rickshaws put the driver behind the passengers. In the Philippines, passengers sat next to the driver in a sidecar.

Cycle-rickshaws were faster than hand-pulled rickshaws. The rickshaw puller became a driver, making passengers less concerned about the morality of having another person pulling them through the streets.[2] Eventually some drivers added electric or gasoline engines to create a motorized or auto-rickshaw.

In the United States, cycle-rickshaws are known as pedicabs. First appearing in Portland, Oregon, in the 1930s, they did not become popular until the 1962 Seattle World's Fair, where a fleet of 20 imported

A pedicab driver waiting for a customer in Barcelona, Spain.

Craig Sunter

Asian cycle-rickshaws carried passengers around the fair. Although the pedicabs had been brought in mostly for their novelty, they turned out to be surprisingly popular and profitable. The vehicle got widespread attention when Elvis Presley rode in one in the 1963 movie *It Happened at the World's Fair.*[3]

Modern pedicabs—made of light materials and more streamlined and comfortable for driver and passengers—remain a primary form of transportation in Asia, India, and Africa. They may be called trishaws, sidecars, pedicabs, cyclos, becaks, samlors, or one of many other local names.[4] They are well suited to narrow streets and can weave through heavy city traffic faster than an automobile or taxi. In the U.S., pedicabs remain a novelty, although their use is growing in most major cities and some smaller tourist destinations. In addition to regular transportation, pedicabs are used for giving city tours and as an alternative to limousines for weddings and proms. Cities have developed local ordinances to govern pedicab operation, licensing, and safety.

Pedicabs are enjoying increased recognition as an efficient, sustainable, green mode of transportation, able to reduce urban CO_2 emissions and traffic congestion. In late 2008, the Centre for Scientific and Industrial Research in New Delhi, India, unveiled the "soleckshaw" (short for solar electric rickshaw), a motorized cycle-rickshaw that runs on a 36-volt solar battery. The soleckshaw can reach speeds of up to 9.3 miles (15 km) per hour, and its battery can run for 30 to 42 miles (48 to 68 km) between charges. New Delhi has set up centralized solar-powered recharging stations where drivers can get their batteries recharged for a small fee.[5] As concern for the environment grows, we may see many more pedicabs on streets throughout the world.

A pedicab at the 1962 World's Fair in Seattle.

Seattle Municipal Archives

Rio de Janeiro or Bust!

Liang Sen/Xinhua/Alamy Live News

Chen Guanming and his rickshaw in Vancouver, Canada, in 2014

In 2010, a Chinese farmer named Chen Guanming left his home in eastern China and began pedaling his cycle-rickshaw to London for the 2012 Olympics. The journey of 37,500 miles took him through 16 countries and more than 1,700 cities. Mr. Chen said he got the idea for his journey to promote the Olympics after attending the Beijing Olympic Games. He slept in his rickshaw while on the road and did odd jobs and gave rides to earn money along the way.[6] His next goal is to attend the 2016 Olympics in Rio de Janeiro, Brazil. Friends arranged his transport with rickshaw from Britain to Halifax, Nova Scotia, where he started this new journey. As of September 2015, he had traveled west from Halifax to Bellevue, Washington, south into California, and then east across the U.S. to New York City. From there he planned to visit Philadelphia and Washington, D.C. before turning south toward Rio[7]—hardly the most direct route! To this traveler, apparently, the journey matters more than the destination.

Skateboards
Have Board, Will Travel

The skateboard started out as just a board with roller-skate wheels attached. Nobody knows who first made one—perhaps a couple of kids looking for something new to do—or exactly when—probably in the late 1940s or early 1950s.

Skateboards would probably have remained a homemade toy if surfers in California hadn't discovered that balancing on a skateboard and "sidewalk surfing" was a good way to hone their skills when they couldn't be riding waves. In 1963, Larry Stevenson, the West Coast publisher of *Surf Guide*, created a professional skateboard based on the shape of a surfboard,[1] and that same basic shape continues today.

Stevenson's board was an improvement, but it still rolled on roller-skate wheels, which were then made of clay. The ride was bumpy at best, and an irregularity in the pavement could cause the wheels to lock, throwing the rider off. Since riders often rode barefoot with no safety gear, it's not surprising that skateboarding was considered dangerous.

In 1972, Frank Nasworthy invented urethane skateboard wheels similar to the ones most skateboarders use today.[2] These new wheels brought a huge improvement in traction and performance and made skateboards much easier to control. Nasworthy named his company Cadillac Wheels to reflect the smoothness of the new ride.

Today's skateboards have come a long way from the early planks on roller-skate wheels, yet they still feature the same basic components: a deck, trucks, and wheels. All of these can be customized to fit the rider and the use for which the board is intended.

There are two basic types of skateboard deck: long boards and short boards. The former give a smoother ride, while short boards are lighter and better for performing tricks and jumps. Decks also vary in width—wider for greater stability, narrower for trick riding—and depth, which is the vertical curvature in the middle of the board. Long boards usually have little or no deck depth, while short boards can be shallower for durability or deeper for more responsiveness when jumping.[3] The tops of most boards are covered with grip tape, a sandpaper-like surface making it easier for the rider's feet to grip the board.

Two trucks—one in the front, one in the rear, each consisting of a baseplate and a hanger—connect the wheels to the deck. The baseplate is screwed to the underside of the deck, and the hanger attaches to the

A skateboard float in the 2011 Quarnevalen parade in Stockholm, Sweden.

Ale Wi

baseplate. A wheel axle runs through the hanger. Trucks help keep the board stable and make it easier to turn.

Skateboard wheels are made of polyurethane, and their size and hardness influence the skateboard's speed and maneuverability.[4]

The popularity of skateboarding has ebbed and

Boys skateboarding and cycling on the Avenida Paulista, São Paulo, Portugal.

flowed over the past 60 years. It has frequently been viewed as an extreme sport or a counterculture activity. Many cities and towns prohibit skateboarding on streets, sidewalks, parking lots, and other public areas, restricting the activity to specially built skate parks. Yet for many people, skateboarding is a cheap, efficient, and environmentally friendly mode of everyday travel, especially in urban areas and on college campuses.[5] Plus it's great exercise!

Go Skateboarding Day

Since 2004, skateboarders worldwide have celebrated June 21 as Go Skateboarding Day. According to the International Association of Skateboard Companies (IASC):

Skateboarders everywhere will show their love and support for skateboarding by holding fundraisers, contests, protests and demos. They'll skate across cities, gather in skate parks, stream into their local skate shop and some will even revel in the solitary act of skateboarding alone at their favorite spot, all bringing together the skateboarding community in the grind heard around the world.[6]

Moving for Fun

Musée des Beaux-Arts de Bordeaux

A Jean Louis Gintrac (1808–1886) painting of shepherds on
stilts in the Landes region of France.

Skateboards aren't the only vehicles that hover between practical use and play. Greeks walked on stilts as early as the sixth century BC, and although stilts can be useful for crossing flooded terrain or spotting lost sheep at a distance,[7] they are also very entertaining toys. Roller skates were invented in 1760 and became wildly popular in the 1860s, when "rinkomania" spread from Massachusetts all the way to Europe.[8] The pogo stick, invented in 1920,[9] is not very practical for getting from point A to point B, but it's great for doing backflips and other daredevil stunts.

Skateboards are like kick scooters without handlebars. After World War II, middle-class children with more time than money could make their own scooters with old roller-skate wheels and scraps of wood and go whizzing around their neighborhoods. Like serious vehicles, scooters and skateboards can expand a person's range—but like pogo sticks, they're mostly just (extreme) fun.

Hovercraft
Riding a Cushion of Air

F lying carpets float on air, but only in cartoons and fairy tales. Hovercraft do it in real life.

A hovercraft rides a cushion of air that supports it slightly above the surface it's traversing—usually a water body, but sometimes ground, snow, ice, swamp, or sand. Powerful downward-directed fans or jets supply the air cushion.

If you've ever tried to walk through hip-deep water, you know that water—which is 815 times denser than air—offers high resistance to any object passing through it.[1] As a result, it takes a big increase in power to achieve even a small increase in the speed of a boat, be it a rowboat or a megayacht. But if a boat can skim *on* the water rather than plowing *through* it—i.e., if it has a *planing hull* (like a high-speed powerboat) rather than a *displacement hull* (like a tugboat or deep-sea fishing boat)—it escapes the tyranny of water resistance and can go much faster. The newest America's Cup multihulled sailboats can lift their hulls completely out of the water at high speeds, riding on thin foils that offer minimal water resistance, and they sail like greased lightning. The final step, of course, is to remove contact with the water altogether, and that brings us to hovercraft.

Although the concept of an air-cushioned vehicle is more than two centuries old, nineteenth-century inventors were unable to design a workable model because they needed more concentrated power than steam engines could provide. They were also unable to keep an air cushion from escaping laterally—that is, off to the sides. Two twentieth-century inventions, the internal combustion engine and the airplane, helped solve these problems.

The SR.N4 Hovercraft (Mountbatten Class) on its last day of service, October 1, 2000.

Andrew Berridge

A gasoline engine could provide the power that a successful air-cushioned vehicle needed, and experience with early airplanes showed that a plane flying at low altitude generates additional lift under its wings, requiring less power to stay aloft. (This is called the *ground effect*.) But it wasn't until 1955 that a British inventor, Sir Christopher Cockerell, put all the pieces together.[2]

Cockerell believed that air pumped into a narrow tunnel around the perimeter of the bottom of a vehicle would flow toward the center, creating an air cushion, and he tested the theory by putting a cat-food can inside a coffee can on a kitchen scale. He used a hair dryer to blow air into the space between the walls of the two cans, and as predicted, the cat-food can rose on the cushion of air when the air pressure measured by the scale equaled the can's weight.

Cockerell determined that making the bottom of the hull concave (curving downward from the center to the edges) and angling the air jets toward the center would help maintain a constant air cushion underneath the vehicle.[3] He launched a working vehicle, which he called a hovercraft, on July 25, 1959, and crossed the English Channel in it. The trip took two hours.[4]

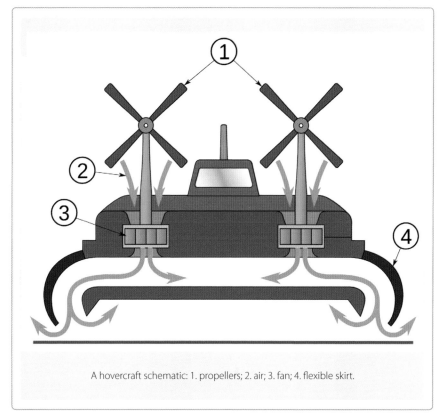

A hovercraft schematic: 1. propellers; 2. air; 3. fan; 4. flexible skirt.

Cockerell's hovercraft did not work well in waves, but another British inventor, C.H. Latimer-Needham, solved that problem by adding a flexible rubber skirt around the outside of the hull to contain the air cushion. This improvement enabled Cockerell's vessel to handle waves of 6 to 7 feet (2 meters), cross uneven marshland, and clear obstacles over 3 feet (1 meter) high. The skirt also increased the air pressure under the hovercraft, allowing it to carry heavier loads with no increase in power.[5]

By the end of the 1960s, hovercraft were providing ferry service on the English Channel between France and Great Britain. Increased fuel prices in the 1970s led companies to switch to more efficient vehicles, however, and the opening of the tunnel under the English Channel in 1994 ended the service. Today, hovercraft are used for search and rescue, surveying, environmental projects, leisure, sport, and racing,[6] and hovercraft enthusiasts have formed clubs in the U.S., Britain, and other countries.

Electric and Hybrid Automobiles
So Nineteenth Century

Electric cars aren't a new idea; they predate gasoline automobiles by half a century. In 1828, Hungarian inventor Ányos Jedlik built a small-scale model car powered by an electric motor. Robert Anderson, a Scot, built the first electric-powered carriage in the 1830s,[1] and other inventors experimented with electric-powered vehicles at the time.

The batteries on which those early vehicles relied were heavy, expensive, and needed frequent recharging, so the vehicles were underpowered, slow, and had limited ranges. By the late 1800s, however, improvements in battery technology led to extensive development of electric vehicles in France, Britain, and the United States. In 1897, the city of New York introduced a fleet of electric taxis (Chapter 18) built by the Electric Carriage and Wagon Company of Philadelphia, and electric cars outsold steam-and gasoline-powered cars in the U.S. in 1900.[2] Then demand plummeted as lower gas prices, improvements in gas-powered engines, and the demand for cars that could make longer trips worked in favor of gasoline-powered cars. Thanks to mass production, the cost of a gasoline-powered automobile fell far below that of an electric one in the 1920s. By 1935, electric vehicles had all but disappeared.

While electric cars rely completely on power from rechargeable batteries, *hybrids* combine the best features of gasoline and electric engines. These are not a new idea, either; two hybrid vehicles were exhibited

This Baker Electrics advertisement from the October 19, 1913 *Washington Post* touted revolving front seats.

at the Paris Salon in 1899. Hybrid technology was being used for commercial vehicles by 1916, and two U.S. electric-vehicle manufacturers, the Baker Company of Cleveland and Chicago's Woods Auto Company, were selling hybrid cars that year. The Woods company claimed its car could reach 35 miles (56 km) per hour and travel 45 miles (72 km) on a gallon of gas, but the car did not sell well.[3] Hybrids, like electric cars, could not compete with cheaper, more powerful gasoline-powered vehicles.

It wasn't until the 1960s that concerns about exhaust emissions and the rising price of gasoline revived an interest in electric and hybrid cars. New government requirements for fuel efficiency, limits on CO_2 emissions, and an increased availability of research funding spurred development of alternative power sources for automobiles in the 1970s.

Electric cars use rechargeable batteries to power an electric motor. When the battery loses its charge, the driver must stop and plug the car in to recharge it. Battery life, the availability of charging stations, and the time required to charge the battery limit the practicality of these cars. Car manufacturers are extending battery life and reducing charging times, however, and as electric cars become more popular, public charging stations are becoming more widely available. In 2007, an Israeli company called A Better Place pioneered the idea of battery

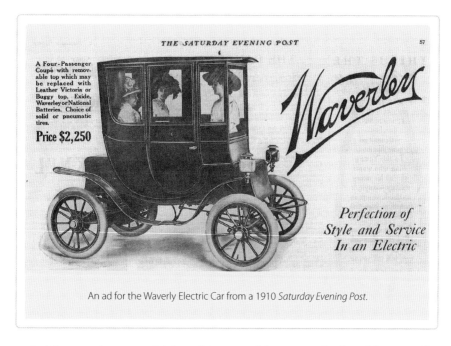

An ad for the Waverly Electric Car from a 1910 *Saturday Evening Post*.

switching stations at which a driver could swap a depleted battery for a fully charged one. Unable to convince automakers to incorporate the technology needed to support battery swapping, the company failed in 2013, but Tesla, a major electric vehicle manufacturer, has been considering a similar system for owners of its cars.[4] The idea is to make recharging as quick and easy as pumping a tank of gas.

The gasoline and electric motors work together in a hybrid vehicle. The combustion engine provides power at cruising speeds while charging the car's batteries, but shuts down and lets the electric motor power the car when it is stopped or moving slowly. Both motors kick in when the demand for power

A plug-in Toyota Prius at a charging station.

is high, such as when the car is accelerating, and a complex array of computerized controllers governs the interplay between the two engines and monitors battery power.[5] Some hybrids use alternative combustion fuels such as diesel, liquefied petroleum (LP) gas, or biofuels.

How Green Are Electric and Hybrid Vehicles?

Electric and hybrid vehicles produce low or no emissions and are highly fuel efficient, but recently some engineers have begun to question whether these vehicles are really as environmentally friendly as is popularly believed. It takes a great deal of energy to produce the lightweight materials used in the cars, and mining the minerals used in batteries and other components depletes scarce rare earth metals and produces toxic wastes that can endanger workers and pollute groundwater. Finally, the electricity used to charge an electric car's batteries may come from a coal-fired power plant.[6] A full inventory of environmental costs and benefits must weigh all factors.

Silence Isn't Always Golden.

One unexpected problem with electric cars is that they make no noise at slow speeds to alert unwary pedestrians. A 2011 study found that electric cars are twice as likely as cars with gasoline engines to hit pedestrians when backing up, slowing or stopping, starting in traffic, or entering or leaving a parking space.[7] People who are partially or completely blind are at particular risk because they rely on sound to determine when it's safe to cross a street. Electric car manufacturers have been asked to add noise to their cars to warn pedestrians of their approach.

All-Terrain Vehicles
A Ticket to the Back Country

The first all-terrain vehicles (ATVs) built in the early 1960s looked a lot like fiberglass canoes or bathtubs on six wheels. Wheel-steered, with room for passengers, these early ATVs could float, and their spinning wheels propelled them short distances across ponds, streams, and swamps. Today, these vehicles are known as *amphibious all-terrain vehicles*, or AATVs.[1]

Today's ATV looks more like a three- or four-wheeled motorcycle riding on large, low-pressure tires, with the third and fourth tires providing more stability than a motorcycle possesses. The rider straddles the seat and steers using handlebars. Most modern ATVs are built to carry one person, although some are two-seat, or tandem, models.

Honda, a Japanese motorcycle company, built the first modern ATVs in 1970 in response to a request from American dealers for a vehicle they could sell in the winter, when motorcycle sales were slow. Honda engineer Osamu Takeuchi took charge of the project and concluded after evaluating a variety of designs that a three-wheeled vehicle provided the best combination of good traction on slippery surfaces with high maneuverability. On 22-inch low-pressure balloon tires, the vehicle he designed traveled easily over rough terrain.[2]

Other motorcycle companies followed suit, and in 1982, Suzuki introduced the first four-wheeled ATV. Three-wheeled models, though light and fast, had proved prone to tipping over in tight turns and could somersault backwards when climbing steep hills because the single-wheeled front end was so much lighter than the rear.[3] Four wheels provided enhanced stability.

ATVs were increasingly popular by the mid-1980s, and the number of accidents and injuries grew accordingly, especially among young riders. The U.S. Consumer Product Safety Commission (CPSC) investigated ATV accidents and safety issues, and their 1986 report suggested that most accidents were due to improper driving, a failure to follow manufacturers' safety warnings, and use of three-wheeled ATVs. In April 1988, ATV distributors in the U.S. agreed to stop selling three-wheeled ATVs and to invest $100 million in safety programs and free

Motorcyclist and quad bike stunt rider Jason Smyth.

Firefighters flanking a forest fire with an ATV fire truck.

U.S. Forest Service, Southwestern Region, Kaibab National Forest

training for ATV purchasers.[4] Although the agreement expired in 1997, few manufacturers resumed selling three-wheeled ATVs.[5]

ATVs were originally envisioned as recreational vehicles for trail riding and hunting, but commercial applications soon developed, and today manufacturers produce utility ATVs for farming, ranching, and other work. Some models include a cargo bed similar to that of a pickup truck; some can pull a trailer. ATVs also enable people with disabilities to enjoy outdoor activities.

Safety is still an issue. A study reported in the *American Journal of Surgery* showed an increase in the number of ATV-injured patients admitted to level-I trauma centers from 164 in 1985–1999 to 269 in 2000–2005. While the number of long-bone fractures increased, however, a decrease in head trauma suggested that state helmet laws have had a positive impact.[6]

Environmental Impacts of ATVs

Increased ATV use has led to greater demand for riding areas. Most states restrict ATVs to marked trails, forbid their use on public roads except to cross them, and set age and training requirements, but not all riders comply with the law. ATV use outside authorized areas has damaged vegetation, increased erosion, disrupted wildlife, damaged sensitive wetland habitats and fish spawning areas, generated silt in lakes and streams, and impacted water supply resources.[7] Local and national ATV organizations discourage off-trail riding and encourage responsible ATV use.

Manned Spacecraft
Out of This World

Johannes Kepler (1571–1630) was a contemporary of the great astronomers Galileo Galilei and Tycho Brahe and a guiding light for the scientific revolution of the seventeenth century. In 1595, he published the first defense of the Copernican solar system (which placed the sun rather than the earth at its center). He took a break from the cosmos in 1611 (while Shakespeare's *Macbeth* was being staged in London) to write the first description of the hexagonal symmetry of snowflakes, and possibly in that same year, he wrote what might be the world's first work of science fiction, a story about a journey to the moon. Called *Somnium* (*The Dream*) and published posthumously, it may have contributed to Kepler's mother—a healer and herbalist—being imprisoned for witchcraft nine years later.

In his 1865 novel *From the Earth to the Moon*, Jules Verne imagined that shooting three people (including a French poet) in a capsule from an enormous cannon could provide enough momentum to escape Earth's atmosphere and reach the moon. In 1903, Russian schoolteacher Konstantin Tsiokovsky published a paper called "The Exploration of Space with Reaction-Propelled Devices" in which he concluded that the force a cannon would require to launch someone into space would be lethal. As an alternative, he designed a multistage rocket and suggested a fuel of liquid hydrogen and liquid oxygen.[1]

In 1909, American college professor Robert Goddard calculated the velocity needed to reach earth orbit, and in 1923, Hermann Oberth, a German teacher, published plans for a two-stage liquid-propellant rocket. That same year, Moscow hosted the world's first exhibition on the prospect of interplanetary travel.[2]

Interest in space exploration increased in the Cold War years following World War II. On October 4, 1957, the Soviet Union launched *Sputnik 1*, the first satellite to reach space, and the following month, *Sputnik 2* carried a dog into orbit and back to earth. The achievement struck

NASA

Astronaut James B. Irwin salutes a U.S. flag during the Apollo 15 lunar surface extravehicular activity (EVA) at the Hadley-Apennine landing site in 1971. This photograph was taken by Astronaut David R. Scott, Apollo 15 commander.

Cosmonaut Yury I. Onufrienko (left) and astronaut Carl E. Walz—Expedition Four mission commander and flight engineer, respectively—wear Russian Sokol suits in the Soyuz 3 spacecraft while docked to the International Space Station (ISS) in 2002.

NASA

Americans with the force of a supernova; if the Soviet Union could look down on the U.S. and its allies from space, what else could it do? The Space Race was on.

American engineers had been developing supersonic flight and an experimental rocket plane, the X-15, which could reach the outer edge of the atmosphere. Some scientists thought the X-15 could be the basis of a new type of plane that would fly into orbit and back to earth.[3] After Sputnik, however, the National Aeronautics and Space Administration (NASA) focused on building a vehicle that could safely lift a human being into earth orbit.

The Soviets landed the first rocket on the moon in September 1959, and on April 12, 1961, Soviet cosmonaut Yuri Gagarin became the first human in space. His flight lasted less than two hours, just long enough to orbit the earth.[4] A month later, American Alan Shepard made a much shorter space flight, and on May 25, 1961, President John F. Kennedy committed to landing a man on the moon before 1970.[5]

Engineers needed a space capsule that could reach the moon, land safely, then blast off and return to Earth. Space suits had to let astronauts move about on the moon's surface while keeping them safe from contaminants and the moon's atmosphere—which, although very thin, does exist and is known to include sodium, potassium, and other gases. Each component of the moon mission was carefully designed and tested. NASA sent unmanned vehicles to land on the moon, and then a manned mission that achieved moon orbit but didn't land. Finally, on July 20, 1969 (four weeks before a very different event, the Woodstock Music Festival), Neil Armstrong became the first man to walk on the moon, telling the world that he was taking "one small step for man, one giant leap for mankind." Two of the three stages of the Saturn rocket that lifted Armstrong's space capsule into orbit were powered by a mixture of liquid hydrogen and liquid oxygen, the same fuel proposed by amateur enthusiast Tsiokovsky in 1903.[6]

NASA planned nine more moon missions and eventual construction of a permanent moon base. Because of budget cuts, however, only five missions were completed, and NASA's focus shifted back to development of a reusable space shuttle.

Today, many scientists question whether manned space travel is needed. Unmanned missions have mapped the surface of Venus and analyzed the soil and atmosphere of Mars. Robotic vehicles can explore a planet's surface without being affected by toxic environments. Rockets and space capsules are expensive to build for use in a single mission.[7] Will scientists and inventors find a new, more efficient means of interplanetary travel?

Did the U.S. Really Land on the Moon?

A Magyar (Hungarian) commemorative stamp of the Apollo
14 moon landing January 31–February 9, 1971.

Some people still don't think so. For over 40 years, a small group of conspiracy theorists have argued that the moon landing was an elaborate hoax played on the American people and designed to fool the Soviet Union into overestimating U.S. scientific capabilities. They contend that the astronauts were launched into space but never left earth orbit, and that video of the landing was shot in a TV studio.

Proponents of the hoax theory have raised many issues as proof to back up their claims, from the angles of shadows shown in photographs to the contention that there couldn't be any photos at all because film would have melted in the 250° F (121° C) temperature. In February 2001, a Fox TV program, *Conspiracy Theory: Did We Land On the Moon?* laid out the supposed evidence that the landing was a fake.[8] Despite extensive and repeated point-by-point rebuttals from NASA and respected non-NASA scientists, the hoax theory continues to this day.

Unmanned Exploration with Remotely Operated Vehicles

NASA

An artist's rendering of one of the NASA Mars Exploration Rovers on the surface of Mars.

More and more scientists are relying on remotely operated vehicles (ROVs) to explore hostile environments and carry out dangerous tasks. These unmanned vehicles have been adapted to a wide range of purposes.

Underwater ROVs are often connected to a ship by a cable that carries both power and commands. Equipped with video cameras and lights, these vehicles can also carry sonars, magnetometers, a manipulator or cutting arm, water samplers, and instruments that measure water clarity, temperature, and density. Underwater ROVs are used to help construct and maintain underwater oil platforms, locate and explore ancient shipwrecks, and study deep-sea marine life.[9] A remotely controlled deep-sea vehicle, *Argo*, discovered the wreck of the *Titanic* in 1985.

Deep space is an even more hostile and challenging environment. ROVs can move about the surface of a planet to collect data on the chemical makeup of the atmosphere and soil and send it back to Earth. Two ROVs on the surface of Mars, *Opportunity* and *Curiosity*, have sent back volumes of important data on the Martian climate and geology, the possible presence (now or in the past) of water, and the planet's habitability. ROVs can also remain on a planet's surface far longer than humans; *Opportunity* has been exploring Mars since 2003.[10]

Jetpacks
Strap One On and Fly

If jetpacks sound like science fiction, it's because for a long time that's exactly what they were. The first mainstream depiction of a jetpack was on the cover of the August 1928 issue of *Amazing Stories*. That same issue contained the first Buck Rogers story, set in the year 2419.

The jetpack on the cover of *Amazing Stories* belonged to Richard Seaton, a character in a story called "The Skylark of Space," and was powered by the mysterious, newly discovered element "X." The wearer was able to hover 100 feet (30 meters) above the ground, move in any direction he faced, fly in loops and figure eights, and land effortlessly. Buck Rogers, featured in his own story in the same issue, flew using power from Inerton, described as the densest substance in the universe.[1] Clearly anyone could have a personal jetpack as soon as someone discovered element "X" or Inerton.

Jetpacks are not just a product of science fiction, however. The Nazis reportedly developed a working jetpack during World War II that enabled the wearer to jump a horizontal distance of 180 feet (55 meters) over minefields, trenches, or barbed wire. No designs, photographs, or prototypes remain, but a working model is rumored to have been delivered to Bell Aerosystems in the U.S. after the war.[2]

On April 20, 1961, an American named Harold Graham used a Rocket Belt developed for Bell Aerospace by Wendell Moore to fly for

Martin Jetpac

The first public flight of the Martin Jetpack at AirVenture 2008 in Oshkosh, Wisconsin.

14 seconds. In what is considered the first successful untethered jet-pack flight, Graham reached a height of 4 feet (1.2 meters) and flew a distance of 35 feet (11 meters) at about 6 miles (9.6 km) per hour.[3] In 1962, after further testing and development, Moore's jetpack reached a top speed of 60 miles (97 km) per hour, a maximum altitude of 60 feet (18 meters), and a flying time of 21 seconds. It also generated 130 decibels of noise, equivalent to the sound of a jet taking off 200 feet (61 meters) away. The biggest risk from Moore's jetpack may have been hearing loss! The belt weighed 100 pounds (45 kg).[4]

By 1962, NASA was developing a jetpack to enable astronauts to maneuver in space when outside their spacecraft. Their model had 10 nozzles and used nitrogen-pressurized hydrogen peroxide jets.[5] In 1965, the movie *Thunderball* showed James Bond using a jetpack to fight evil villains. The TV show *Lost in Space* and articles in *Popular Science* magazine continued to popularize the idea of the jetpack, and Bell Aerospace continued developing their Rocket Belt.

In 1984, NASA used an updated version of the jetpack—called a manned maneuvering unit (MMU)—on three space shuttle missions. But jetpacks designed for use on earth were still heavy, noisy, difficult to control, and could fly less than a minute. The 1990s did not yield significant improvements.

The most recent effort to develop a working jetpack is the Martin Jetpack, developed by the Martin Aircraft Company of Christchurch, New Zealand, and introduced in 2008. Martin advertises its jetpack as "the world's first practical jetpack, set to revolutionise the industries of aviation, recreation, and transportation."[6] Martin envisions early use of its jetpack as a first responder vehicle, after which it plans to develop models for leisure and personal use. In 2013, Martin's prototype was approved for manned test flights by the New Zealand Civil Aviation Authority. Critics point out that the Martin Jetpack is still too heavy, noisy, and expensive for personal use.[7]

The Flying Wing

In November 2006, Swiss pilot Yves Rossey tested a "flying wing" that let him fly for 6 minutes, 9 seconds. His apparatus—which he steered with body movements—combined four kerosene-fueled jet engines with a rigid deployable carbon-fiber wing having a 3-meter (9.8-foot) span. Rossey has continued to test and demonstrate his vehicle, descending from the Alps, flying over the Grand Canyon and the English Channel, and dropping from a hot air balloon at 7,900 feet (2,409 meters). His wing is not technically a jetpack, since it must be dropped from a height, but his goal is to add ground take-off to its capabilities.[8]

Hydro-Jetpacks

Kevork Djansezian / Getty Images/Thinkstock

Flying with the Jetlev hydro-jetpack in Newport Beach harbor, California, 2012.

One form of jetpack is available for public use and gaining popularity: the water-powered or hydro-jetpack. This device works only over water and propels itself by sucking up water through a hose and shooting it downward through nozzles. The thrust from the nozzles keeps the rider in the air, as anyone who's ever felt the recoil from a high-velocity fire hose can readily believe. Currently these jetpacks are available in a limited number of places, usually warm-water tourist destinations, but their numbers are expected to increase.

Bullet Trains
High-Speed Rail

I n 1879, German engineer Walter von Siemens demonstrated the first working electric rail locomotive. His miniature locomotive pulled 30 people using a 3-horsepower engine drawing current from a third rail in the track.[1] Electric locomotives were well suited for mines and subways because they required no combustion to cause fires and generated no smoke to pollute underground tunnels (Chapter 31). In 1907, the New York, New Haven & Hartford line in the U.S. became the first mainline railroad to electrify.[2] Electric trains could reach speeds of 60 miles (96 km) per hour.

Electric engines were reliable, but converting railroads to electricity was expensive. Every inch of track, including rail yards and switching stations, had to be electrified unless steam-powered locomotives were provided to move trains through nonelectrified segments. There were also questions about whether an electric engine that was light enough to be practical could produce enough power to pull heavy freight.

Diesel engines seemed to provide a perfect power solution for trains. In 1930, General Electric developed a compact, powerful two-cylinder diesel engine, and in 1933, the Union Pacific and Chicago, Burlington & Quincy railroads built diesel-powered trains.[3] Diesel locomotives were faster, more reliable, and cheaper to operate than steam locomotives, and they soon became the new standard in the U.S.

The United Kingdom and much of Europe, however, continued to develop electric railways. Bombing during World War II destroyed major portions of the railway infrastructure in these countries, and improvements in electric locomotive technology made electricity an attractive option when rebuilding.

The Japanese bullet train Shinkansen 500, seen in 2010.

By the 1960s, railroads were facing competition as a preferred means of travel. Airplanes had become safe and affordable, and traveling by plane was fast. More people owned cars, and highway systems linked major cities, making travel by automobile a good option. Railroads would need to become faster and more efficient if they wanted to stay in the passenger business.

The French were the first to experiment with high-speed rail travel. In 1955, France tested two 4,000-horsepower trains that reached a record speed of 205.6 miles (329 km) per hour.[4] It was the Japanese, however, who proved that day-to-day operation at more than 100 miles (160 km) per hour was possible. They introduced the first of the Shinkansen ("New Railways") trains on the New Tokaido line in October 1964, and five years later, eight trains were traveling each way on that line, carrying more than half a million people annually at speeds averaging 101 miles (162 km) per hour.[5] The Shinkansen trains came to be known as bullet trains because of the shape of their locomotives. Today a network of high-speed train lines serves Japan's main islands, connecting Tokyo with most of the country's major cities. Trains run at speeds of up to 198 miles (320 km) per hour on lines exclusively built for and used by Shinkansen trains.[6]

Many countries have invested in the resources needed for high-speed rail travel. High-speed trains in China run at an operational speed of 155 to 186 miles (250 to 300 km) per hour, but can reach speeds over 248 miles (400 km) per hour. A bullet train can take a passenger from Beijing to Gangzhou in eight hours, a trip that would take 24 hours on a conventional train.[7] The European Union has defined standards for high-speed rail and is working to link member countries' railways into a network for travel and shipment of goods throughout the continent.[8]

The United States has been slow to embrace high-speed rail. A report prepared for Congress in 2013 cited lack of long-term funding as the major obstacle. Accommodating high-speed trains would require upgrading existing tracks or building new ones, both expensive options. Without a long-term commitment from the federal government to help pay construction and operating costs, states are reluctant to proceed with high-speed rail projects.[9]

Mag-lev: Faster Than a Speeding Bullet Train

Maryland GovPics

An L0 series mag-lev train on the Chuo Shinkansen test track in Japan.

Some countries are developing high-speed electromagnetic trains that float in the air over a guideway.

The basic principle behind electromagnetic propulsion is that opposite poles attract and like poles repel each other. In other words, two magnets with the positive poles facing each other will push themselves apart.

Mag-lev (magnetic levitation) trains have three components: a large electrical power source; metal coils that line a guideway or track; and large guidance magnets attached to the underside of the train.[10] The magnetized metal coils in the guideway repel the magnets mounted under the train, causing the train to levitate above the track. With no friction to slow it down, a mag-lev train can reach speeds as high as 310 miles (500 km) per hour. Electrical power running to the coils in the guideway creates alternating magnetic fields that regularly change the polarity of the coils. This causes the magnetic field in front of the train to pull it forward, while the magnetic field behind the train pushes to add more thrust.[11]

The first commercial mag-lev train began regular service in China in December 2003. It can go 19 miles (30.5 km) in approximately 10 minutes. In Japan, a new mag-lev railway is scheduled to open in 2027, and its builders claim that it will cover the distance from Tokyo to Nagoya in less than half the time of the high-speed bullet trains now in use.[12]

Mag-lev trains are said to be cheaper, safer, quieter, and better for the environment,[13] but critics point out that they require expensive construction of dedicated guideways. Others worry that exposure to magnetic fields could be dangerous for passengers, but developers claim that mag-lev field strengths are below hazardous levels.[14]

Supersonic Airliners
Breaking the Sound Barrier

On October 14, 1947, Captain Charles "Chuck" Yeager, an accomplished World War II fighter pilot, became the first pilot to "break the sound barrier" and fly faster than the speed of sound. Yeager was flying the X-1 *Glamorous Glennis,* an experimental plane developed by the Bell Aircraft Company and powered by a four-chamber rocket engine that had been built to gather information on aerodynamic conditions close to the speed of sound.[1] Today, *Glamorous Glennis*, its fuselage shaped like a .50-caliber bullet, hangs on display in the Smithsonian Air and Space Museum in Washington, DC.

The history of airplane development was fueled by the drive to fly farther and faster. In wartime, faster, more maneuverable planes helped their pilots win dogfights and evade antiaircraft fire. In times of peace, airlines that could reach their destinations faster could make more round trips, increasing profits. No one doubted that supersonic flight would be the next big accomplishment.

Yeager's success led to more sophisticated experimental supersonic planes. The North American X-15, which first flew in September 1959, reached speeds of up to Mach 6.72 and altitudes of 354,200 feet (67 miles or 108 km), the outer limits of Earth's atmosphere and the beginning of space. Eight pilots earned their astronaut wings flying the X-15.[2]

US. Air Force

Left to right: Chuck Yeager, Gus Lundquist, and Jim Fitzgerald (wearing a flight suit) stand next to the Bell X-1 rocket research airplane *Glamorous Glennis*.

Experimental test planes were lightweight with high-powered engines and carried only one or two people. The challenge for commercial airlines was to design a supersonic airliner that was large enough to carry passengers comfortably. The new plane would not fly as high or as fast as the X-15—and sound travels faster at sea level than at high altitude—but it would still be fast enough to break the sound barrier.

In December 1968, the Soviet Union debuted the Tu-144, the first civil aircraft to reach Mach 2. The plane appeared at the Paris Air Show in 1973, where it crashed at the end of its performance, killing the crew and eight spectators.[3] Meanwhile, the British Air Ministry was partnering with a French firm to develop their own supersonic plane, the Concorde, which crossed the Atlantic for the first time in September 1973.

British and French airlines had assumed that a supersonic plane that could cross the Atlantic in three and a half hours would be a passenger's first choice. Unfortunately, the Concorde consumed far more fuel than a subsonic jet aircraft of similar capacity, making it expensive to operate. To break even in 2000, the airline needed to charge $9,000 (£6,000) for a round-trip ticket,[4] and only a limited number of passengers were able and willing to pay such a price.

The Concorde was also much noisier on takeoff and landing than a subsonic passenger jet, and its supersonic flight caused a trailing *sonic boom* (the explosive noise of the shock waves emanating from an object traveling faster than sound). The crack of a supersonic bullet may not be deafening, but the sonic boom from an object the size of the Concorde is something else again. Supersonic flights by the U.S. Air Force were already generating noise complaints, and groups such as the Citizens League Against the Sonic Boom opposed supersonic flight.[5] Some countries, including the U.S., banned the planes from flying over land. Environmental groups such as the National Wildlife Federation and the Sierra Club expressed concern that the planes would damage the upper atmosphere and have harmful effects on humans and wildlife.

The final flight of a Concorde, in 2003.

Adrian Pingstone

The Concorde continued making transatlantic flights until it was retired in 2003, but political opposition and the inability to make supersonic flight profitable deterred development of additional supersonic passenger aircraft. Instead, aircraft manufacturers concentrated on making large, wide-bodied conventional jets such as the Boeing 747 and the McDonnell-Douglas DC-10 to increase the number of passengers that could be carried on each flight.

This NACA (National Advisory Committee for Aeronautics) High-Speed Flight Research Station photograph of the XF-92A was taken around 1953 near Edwards Air Force Base in California.

Mach

Mach is a number indicating the ratio of the speed of an object to the speed of sound in the medium through which the object is moving. For example, an aircraft moving twice as fast as the speed of sound is said to be traveling at Mach 2. The speed of sound at sea level and a temperature of 15° C (59° F) is 761 miles (1,225 km) per hour, or 661.5 knots, At 36,000 feet (11,000 meters) and a temperature of −50° C (−58° F), sound waves travel 15 percent slower, but a plane breaking the sound barrier in either instance would be traveling at Mach 1.

Space Shuttles
Into Space and Back Again

Neil Armstrong's successful moon walk in 1969 (Chapter 45) demonstrated that humankind could reach an off-planet destination—even if the moon is little more than the front stoop in our journey from Earth into the cosmos. But moon missions were expensive, in part because they required construction of large booster rockets that were used once and then burned up in Earth's atmosphere. To make space travel more sustainable, NASA wanted to develop a reusable vehicle that could fly into orbit and land safely back on Earth.

In 1958, the U.S. Air Force had developed an experimental rocket plane, the X-15 (Chapter 48), which could fly to the outer edge of Earth's atmosphere, where space begins. Some engineers, thinking the X-15 could be the basis for a space plane,[1] had begun planning for the X-20 Dynasoar, designed to be launched on a Titan rocket and glide back to earth for a runway landing.[2] But federal budget cuts and pressure to beat the Soviet Union to the moon shifted NASA's focus to designing a space capsule that could be built quickly with existing technology.

By the early 1970s, NASA had renewed its interest in a reusable space vehicle. The Soviets launched the first orbiting space station, Salyut 1, in April 1971. By mid-1974, Salyut 1 had been replaced by Salyut 3, and the first crew had spent time living aboard the station. The U.S. launched its own space station, Skylab, in 1973.[3] With a permanent

NASA

A launching of the space shuttle *Columbia*.

space base available, astronauts needed an affordable, reliable way to travel back and forth, and the old idea of a space plane became current again.

On April 12, 1981, the U.S. launched the space shuttle *Columbia* with two astronauts aboard for its first test in space. NASA soon discovered that the shuttle was more expensive to operate than originally envisioned, however. In addition, the European Space Agency offered companies a less expensive option for deploying satellites or other services,[4] cutting into the revenues NASA had hoped would help fund shuttle operations.

Despite these roadblocks, NASA built five shuttle vehicles and established a schedule of launches. Together, the five shuttles flew more than 130 times, traveled more than a half billion miles, and carried more than 350 people into space.[5] To show that the shuttle was safe, NASA began including civilians in the crews along with astronauts. They hoped the positive publicity would increase support for the program and lead to additional federal funding.

In 1986, the *Challenger*—NASA's second shuttle—exploded on liftoff in its tenth mission, killing all of the crew including elementary school teacher Christa McAuliffe. And in 2003, the *Columbia* disintegrated on reentry into Earth's atmosphere after its twenty-eighth mission, killing all seven crew members.

Inside the International Space Station's Node 1 or Unity, NASA astronaut Chris Ferguson, STS-135 commander, adds his mission's decal as the final piece of the collection of shuttle crew insignias. NASA astronaut Mike Fossum, Expedition 28 flight engineer, looks on.

While the U.S. was developing the space shuttle, the Soviet Union concentrated on building bigger and better space stations. In 1986, they replaced Salyut with Mir, which had more comfortable living quarters and better life-support systems. The subsequent dissolution of the Soviet Union, the end of the Cold War, and consequent cuts to Space Race budgets encouraged the U.S. and Russian Federation space agencies to cooperate, and by 1995, American astronauts were joining Russian cosmonauts living and working on Mir. Eventually, the U.S., Russia, Canada, Japan, and the 11 nations of the European Space Agency created a consortium to build the International Space Station

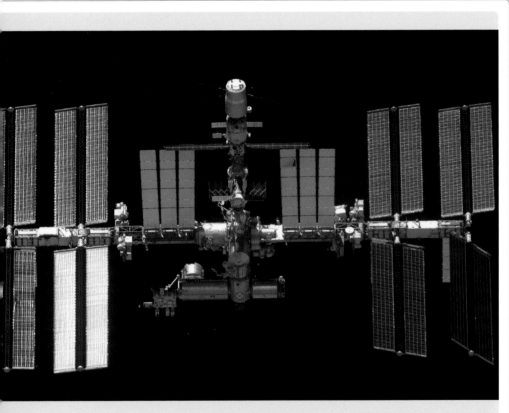

NASA

A view of the International Space Station from the U.S. Space Shuttle *Endeavour* after undocking, May 30, 2011.

(ISS). Space shuttles proved essential for carrying ISS modules into space and joining them with the main body, as well as carrying equipment and supplies to the ISS crew. NASA was designing a second generation of shuttles.

But in 2004, the focus of the American space program changed again. NASA's new priority, as announced by President George W. Bush, would be setting up a permanent moon base and putting astronauts on Mars.[6] Given limited funding, NASA was forced to drop its plans for newer versions of the space shuttle. The last NASA shuttle flight ended on July 21, 2011, with the safe landing of the *Atlantis*.

Commercial Space Flight

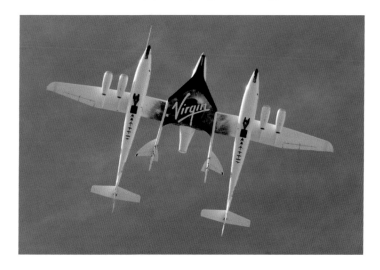

In 1996, the X Foundation offered a $10 million prize for the first person who could reach suborbital space (an altitude of 60 miles or 100 km) twice in the same craft within two weeks. American Burt Rutan won the prize in 2004 with his experimental vehicle *SpaceShipOne*.

(Continued on next page)

(Continued from previous page)

Rutan then joined with British entrepreneur Sir Richard Branson to set up Virgin Galactic, a space tourism company. In December 2009, they introduced *SpaceShipTwo* (SS2). While NASA employed reusable external solid rocket boosters to lift its shuttles off the ground, SS2 was to be carried aloft by a specially built carrier plane, *WhiteKnightTwo*. At approximately 50,000 feet (9.5 miles or 15 km), SS2 would separate from the carrier and fire its own rocket engines to ascend into orbit. The company estimated that after 18 months of testing, the SS2 would be ready to begin taking paying passengers for a trip to the edge of space orbit. Despite the initial $200,000 ticket price and long wait, well-heeled adventurers jumped at the chance for a two-hour flight with six minutes of weightlessness.[7] By late 2014, the ticket price had risen to $250,000, and the number of reservations stood at more than 700.[8]

On October 31, 2014, the SS2 crashed during a test flight, killing one of the two pilots and destroying the vehicle. The flight was the first with a motor powered by a new plastic-based fuel. Virgin Galactic intended to continue moving ahead with its plans, and a second *SpaceShipTwo* was under construction in late 2015.[9]

A few days before the SS2 crash, an unmanned Orbital Sciences Corporation rocket carrying supplies to the International Space Station exploded 15 seconds after launch.[10] The two crashes cast doubt on whether commercial space vehicles will be able to fill the void left by the end of NASA's shuttle program; however, in December 2015, a Falcon 9 rocket developed by SpaceX landed safely back on Earth after launching its payload into orbit, providing new hope that reusable rockets are possible.[11]

The Segway
Pedestrians on Wheels

Segway Inc., in Bedford, New Hampshire, bills itself as the leader in personal green transportation. In December 2001, when company founder Dean Kamen unveiled his new invention, the Segway Personal Transporter (PT), on the popular TV show *Good Morning America*, he described it as "the world's first self-balancing human transporter."[1] Kamen's goal had been to create a vehicle that took up minimal space, was highly maneuverable, and could safely share sidewalks and pathways with pedestrians. The vehicle should be nonpolluting and reduce, rather than add to, traffic congestion in urban areas.[2] The result was an innovative approach to personal transportation.

This two-wheeled, electric-powered vehicle—ridden by one standing passenger—is unique for several reasons. First, it's a zero-emission vehicle powered by a lithium-ion battery that can be recharged by plugging it into a household electrical socket. The company claims that electronic propulsion also enables accurate turning, a smooth ride, and precise traction control and braking. Second, the vehicle uses dynamic stabilization technology for balancing, making it easy for the rider to remain upright.[3] A Segway stands upright even when stopped and turned off. Finally, the rider steers the Segway by leaning in the direction he or she wants to go, without the need for a wheel or handlebar steering.

Although the Segway was developed for personal use, its makers quickly saw additional opportunities. Since 2003, Segway Inc. has developed versions for businesses, police and security agencies, and the U.S. military. Warehouse personnel, airport security guards, and urban police departments all use Segways. At the Beijing 2008 Summer Olympics, more than 100 Segway PTs were used by event and

People on a Segway tour in Yamanashi, Japan.

airport security, arena logistics personnel, and Olympic Ambassadors. Segway Inc. has partnered with the U.S. Defense Advanced Research Projects Agency (DARPA) to convert the PT into a Robotic Mobility

NASA

Will we soon see Segway-riding robot space explorers? Shown here is Robonaut B, one of two NASA robots used in recent hand-in-hand testing with humans at the Johnson Space Center to evaluate their shared ability to perform extravehicular tasks.

Platform, resulting in a new product for the unmanned ground robotics industry. Other special versions of the original Segway PT have been developed for varied terrain (the "off-road" Segway) and golf courses (the Golf Transporter). Tour operators in cities from San Francisco to New York to Berlin and Paris offer Segway tours.

Introduction of the Segway was not without challenges. State laws and city ordinances had to be created, or existing laws modified, to allow the devices to operate on sidewalks. As of June 2014, 44 states and the District of Columbia have enacted legislation to allow use of Segway PTs on sidewalks, bike paths, and certain roads.[4] Because the Segway can reach speeds of over 12 miles (20 km) per hour, government agencies have been asked to consider whether riders should be required to wear helmets. Complaints from pedestrians about Segways monopolizing sidewalk space have caused some cities, such as Chicago, to consider limits on the number of participants in a Segway tour.[5]

There is at least one use for the Segway PT that its inventor probably never foresaw: Segway polo. Developed by members of the Bay Area Segway Enthusiasts Group in 2004, the game is similar to horse polo. Two teams of five players hit a ball with mallets, and points are earned by hitting the ball into the other team's goal. Segway polo teams have formed in a number of countries, including Germany, Barbados, and New Zealand, and a governing body, the International Segway Polo Association, has been established, along with an annual world cup championship match.[6]

What's in a Name?

The name Segway is a homophone (a word that sounds the same as another word) of *segue*, meaning a smooth transition.[7] The name relates to the smooth ride promised by the Segway PT. Indeed, Segway rides are often referred to as "glides."

Conclusion

Jaylin wakes up feeling refreshed after a good night's sleep. As she opens her eyes and stretches, the hatch of her stasis pod opens and she sees her parents and younger brother doing the same. Everyone looks just the way they did when they left Earth orbit four years ago, thanks to the pods' ability to suspend their metabolic functions and slow aging. Even so, Jaylin is thankful that the newest generation of Galactic International spaceships is equipped with faster-than-light hyperdrives; otherwise, the trip would have taken twice as long.

This is the first interplanetary vacation Jaylin's family has ever taken, but they have a good idea what to expect. They took a virtual tour of the resort and the off-site excursions before they booked the trip. Jaylin hopes the in-person experience will be as much fun.

A driverless shuttle meets the family at the spaceport and whisks them off to the hotel. There's no need for luggage; the resort will provide everything they need, based on an electronic survey they completed before leaving Earth and information collected in the stasis pod. The number of spaceships traveling to this newly developed asteroid is still small, and cargo space is reserved for the resort's provisions. Right now Jaylin thinks that's a great idea, because she is starving!

Is Jaylin's trip just science fiction or a vision of our future? It could be both.

Readers of Jules Verne's 1873 novel *Around the World in Eighty Days* were still getting used to the idea that it was possible to travel around the globe in one lifetime, let alone in 80 days. They would have been astounded to hear that space shuttles and bullet trains would someday be a reality. Scientists and science fiction authors have always envisioned new ways to travel long before the technology needed to make them work existed, and it's reasonable to think that will continue. The rocket that took Neil Armstrong to the moon in 1969 used a fuel first suggested by a Russian schoolteacher in 1903. Leonardo da Vinci designed helicopters almost 500 years before someone built one that worked. And da Vinci's design for a human-powered Helical Air Screw was considered unworkable until researchers built one in 2013 using new ultralight materials.

Astronomer Johannes Kepler's 1611 novel *Somnium* (*The Dream*) was one of the earliest to depict a journey to the moon. In *From the Earth to the Moon*, published in 1865, Jules Verne described space flight with light-propelled spaceships; today, scientists are working on the development of solar sails.[1] The submarine *Nautilus* in Verne's *Twenty Thousand Leagues Under the Sea* displays many features of modern vessels despite the fact that the first mechanically powered submarine, the French Navy's *Plongeur*, had been built only seven years earlier.

Some companies recognize the value of imagination. Microsoft, Google, Apple, and others have invited science fiction writers to give talks to employees and meet with research and development teams, becoming part of the product design process. Companies have even commissioned "design fiction," imaginative works that model how people would use new products. The "what-if" story helps companies decide if a new product is worth developing.[2]

Humans have always been a restless species. We may not know what future vehicles will look like or how they will work, but history allows us to make predictions.

* Vehicles will continue to be designed for specific purposes—some serious, such as interplanetary travel; some as simple as having fun. Unmanned vehicles or robots equipped with sensors and video cameras will enable us to "travel" to places too dangerous to explore in person.

* Inventors will continue to refine and build on existing vehicles, often blurring the lines between vehicle types.

Is an amphibious car really a boat? Is a flying submarine really a plane? If a Chinese bullet train can travel at over 248 miles (400 km) per hour today, why not 800 miles (1,287 km) per hour twenty years from now? Entrepreneur Elon Musk, the founder of Tesla Motors, calculates that a magnetic-levitation train traveling in an undersea vacuum tube could reach a speed of 800 miles an hour, making the trip from Manchester, England, to New York City in four hours.[3]

* Vehicles first invented for special purposes will move into common use and become more affordable. Once colonies are established on the moon, Mars, or deep underwater, travel by space shuttle or submarine will become as routine as taking a bus, train, or airplane is today.

* Energy efficiency and environmental concerns will play a greater role in determining which vehicles gain popularity. Automakers and other vehicle manufacturers are already adopting electric and hybrid technologies to reduce carbon-dioxide emissions and increase efficiency. Bicycles and pedicabs are gaining popularity as more environmentally friendly ways to travel. Some inventors are addressing environmental concerns by reviving older technologies with solar or wind power, such as a modern clipper ship with a 180-foot sail and biomethane engines that can carry 4,500 tons of cargo. If a safe enough gas can be found, dirigibles may come back into use as a kind of ocean liner in the sky.[4] And will personal teleporters become the next generation's preferred mode of zero-emissions commuting?

* Leisure travel and the newest, most exotic vehicles will continue to be luxuries. Wealthy young gentlemen of the seventeenth to nineteenth centuries finished off their educations with a Grand Tour of cultural sites throughout Europe. Today Virgin Galactic and other companies are

exploring the possibility of sending civilians to visit space, but only the wealthiest can afford the trip.

* How people travel will continue to reflect their wealth and social class. Ancient Chinese emperors rode in ornate sedan chairs carried on their servants' shoulders. When those same servants had to go somewhere, they walked. Long-distance space voyages of the future may resemble the ocean liners of the late nineteenth century, with first-class passengers traveling in luxury and third-class ticket-holders piled into bunks next to the engine room or working on board to pay for their trip.

Cheaper, faster travel will enable people to take short vacations farther from home, continuing the trend toward globalization. Self-driving vehicles will put taxicab and bus drivers out of work—but will they also create new jobs, and if so, what kind? Virtual travel will open new vistas to people unable to travel physically; anyone will be able to go anywhere and see anything they want. History students will get an in-class virtual tour of the Colosseum rather than a lecture and slideshow. The rise of the automobile in the 1900s led to the growth of suburbs as people no longer needed to live close to their jobs. Routine use of teleportation would expand that trend; people would no longer need to live in the same state as their job, the same country or, perhaps one day, even on the same planet.

A hundred years from now, will walking still be humans' primary method of moving about, or will we rely on a personal transporter that rides on a cushion of air? Will some of us be walking on custom-designed prosthetic legs programmed to respond to our brain's neural impulses? Only one thing is certain: as long as humans endure, we will be finding new places to explore and new ways to get there.

Glossary

ama: See *outrigger*.

barouche: A four-wheeled, horse-drawn carriage, popular in the nineteenth century, with a driver's seat high in front, two double seats facing each other inside, and a collapsible top over the back seat.

biomethane: A naturally occurring gas produced by the breakdown, without oxygen, of organic matter such as sewage, food waste, or grass and other plants, making it a renewable energy source.

bustle: In nineteenth-century fashion, a frame or pad worn just below the waist at the back of a woman's skirt that made the back of the skirt look larger and fuller.

carbon fiber: A high-tech, man-made, very light and strong construction material composed mostly of carbon atoms that has enabled big weight savings in vehicles on land, sea, air, and space.

center of gravity: The imaginary point in an object where its weight is evenly distributed and all sides are in balance. Objects with a lower center of gravity and a wide base are more stable and harder to knock over.

carbon dioxide (CO_2): A naturally occurring colorless, odorless gas, necessary for plants. As the principal end product from the burning of fossil fuels, atmospheric carbon dioxide has increased since the dawn of the Industrial Revolution and is now recognized as a greenhouse gas and a driver of ocean acidification.

crankshaft: A metal shaft that converts the up-and-down travel of an engine's pistons into rotational energy that is transmitted via gearing to the vehicle's wheels, paddles, or propellers.

dirigible: A gas-filled, lighter-than-air craft that, unlike a hot air balloon, is self-propelled and steerable.

emigrant: A person who leaves his or her native country to escape violence, persecution, or lack of opportunity and to seek a better life in another country.

eminent domain: The legal right of a government to take private property for public use—as in a road, subway line, or public housing—even without the consent of the property's owner, following payment of the property's fair market value.

frock coat: A close-fitting man's suit coat, usually double-breasted, that falls below the knee with a slit in the back, popular in the nineteenth century and now worn chiefly on formal occasions.

gig: A lightweight, two-wheeled cart, common before the twentieth century. A gig was pulled by one horse, could seat two people, and had space behind the seat to carry belongings.

hackney: A coach or carriage for hire.

indigenous: Originating or occurring naturally in a place, rather than having arrived from somewhere else.

infrastructure: The basic physical structures and facilities, such as buildings, roads, bridges, rail lines, and power plants, needed for society to function.

knot: A measure of speed used at sea. A ship traveling at one knot would be covering one nautical mile (1.852 km) or 1.151 statute miles per hour.

light-year: The distance traveled by light in one year. Since the speed of light is 186,000 miles (299,792.5 km) per second, a light-year is approximately 5.88 trillion miles (9.46 trillion km).

lorry: The British term for a truck used to carry heavy loads of cargo or military troops.

millennium: A period of one thousand years or ten centuries.

nautical mile: A unit of distance used at sea and equal to one minute of latitude. A nautical mile is 1.151 statute or land miles.

omnibus: A public vehicle built to carry a large number of passengers; a bus.

outrigger: A light, thin secondary hull on a boat, fastened rigidly to and in parallel with the main hull, the purpose of which is to stabilize an otherwise unstable hull. The outrigger of a Polynesian sailing canoe, and of some modern sailboats derived from this type, is called an ama.

papyrus: A tall, grass-like plant that grows in marshes, especially in Egypt. Its fibrous stems were once used for making rope, sandals, boats, and paper.

pedicab: A three-wheeled, pedal-driven vehicle with a covered cab for two people in back of the driver.

perimeter: The outside edges of an object or area.

permafrost: A layer of soil below the surface that is always frozen, even when the ground above it thaws, found mostly in very high latitudes.

pneumatic: Inflated or operating with compressed air.

prototype: An early working model, especially of a machine, that can be tested and reworked as necessary to refine the design.

quarantine: The temporary segregation and isolation of people, animals, or things (such as plants) that might be carrying a disease, to prevent the spread of the disease.

rationing: Limiting the distribution of a commodity that is in short supply.

revenue cutters: Small, fast, lightly armed boats used to prevent smugglers from bringing goods into a country illegally and without paying the required taxes.

sedan chair: An enclosed chair with horizontal poles on either side, carried by two or more people, commonly used in the seventeenth and eighteenth centuries.

sisal: A strong, white fiber from the leaves of the Mexican agave plant, used in a wide range of products including rope, twine, cloth, carpets, and paper.

sledge: A vehicle on runners, like a large sled, used to carry cargo or passengers over snow, ice, or rough ground. Sledges were pulled by horses, dogs, or other work animals.

sprocket: A wheel with a row of teeth around its edge that engages with opposing holes or recesses in a chain or a roll of film, tape, or paper, causing it to turn when the wheel turns.

steppes: Large, flat expanses of grassland with few trees, especially in southeastern Europe and Asia.

tackle: A system of ropes and pulleys (or *blocks*) for lifting, lowering, or moving objects that magnifies force (i.e., confers *mechanical advantage*) by reducing distance traveled. For example, 50 pounds of pull on the hauling end of a two-part tackle through a distance of 12 inches can theoretically lift a 100-pound object by a distance of 6 inches—though the friction of the rope in the pulleys will prevent this theoretical outcome from being achieved in practice.

torque: Rotational power.

trap: A light one-horse carriage that could seat two people.

travois: Two long poles lashed together in a V shape with the pointed end carried and pulled by a person or animal. A platform or netting is fastened between the poles, behind the load hauler, in order to carry goods or family members who are unable to walk. Though crude, travois were effective in mountainous, marshy, or snowy terrain where wheeled vehicles could not have traveled.

tundra: A large, treeless expanse of flat land in the Arctic regions of Europe, Asia, and North America where the subsoil is always frozen.

turboshaft jet engine: A gas turbine engine designed to produce shaft power, such as that needed to turn a helicopter rotor, rather than jet thrust.

two-stroke versus four-stroke (two-cycle versus four-cycle) engine: A reference to the number of times each piston in an internal combustion engine must move up or down (each up or down travel being a "stroke") to create one full revolution of the crankshaft. In a four-stroke engine the four strokes accomplish air intake; air compression; fuel injection/ignition (this being the power stroke); and exhaust. A two-stroke engine reduces these four steps to two, giving it twice as many power strokes and thus more power per unit of engine weight, but this advantage comes with losses of fuel efficiency and durability and greater exhaust pollution.

U-boat: The common name for *Unterseeboot*, or undersea boat, a German submarine. U-boats were deployed in World Wars I and II with devastating effects.

way station: A place to stop for rest, food, or supplies during a long journey between towns or railway stations.

winch: A machine featuring a rotating drum around which a chain or rope is wound. The drum is turned by means of an arm that is longer than the drum diameter, thus conferring a mechanical advantage for lifting or pulling heavy objects (or sails loaded by wind pressure).

yoke: A crossbar with two U-shaped attachments that fit around the necks of a pair of horses or oxen to attach them to a wagon or other vehicle so that they can pull it as a team. Yoking is the act of fastening animals into a yoke.

Sources and Additional Resources

1. Shoes: Feet First!

Connolly, Tom. "The World's Oldest Shoes." University of Oregon: http://pages.uoregon.edu/connolly/FRsandals.htm.

Field, Simon Quellen. "What Are Shoes Made From?" *200 Questions About Chemistry…with Answers* (2011): http://questions.sci-toys.com/node/65.

Gillon, J.K. "Robert Barclay Allardice: The Celebrated Pedestrian": http://gillonj.tripod.com/thecelebratedpedestrian/.

Gugliotta, Guy. "The Great Human Migration: Why humans left their African homeland 80,000 years ago to colonize the world." *Smithsonian Magazine* (July 2008): http://www.smithsonianmag.com/history/the-great-human-migration-13561/?all.

Koerth-Baker, Maggie. "First Shoes Worn 40,000 Years Ago." *LiveScience* (June 5, 2008): http://www.livescience.com/4964-shoes-worn-40-000-years.html.

Ravilious, Kate. "World's Oldest Leather Shoe Found—Stunningly Preserved." *National Geographic News* (June 11, 2010): http://news.nationalgeographic.com/news/2010/06/100609-worlds-oldest-leather-shoe-armenia-science/.

Rosenbaum, Mike. "An Illustrated History of Race Walking." About Sports, About.com: http://trackandfield.about.com/od/distanceevents/ss/illusracewalk.htm.

This Day in History. "Jul 20, 1969: Armstrong Walks on Moon." *History*: http://www.history.com/this-day-in-history/armstrong-walks-on-moon.

Additional Resources

"The Human Journey: Migration Routes." Genographic Project. National Geographic: https://genographic.nationalgeographic.com/human-journey/.

Unit 3: Human Migrations. Bridging World History. Annenberg Learner: http://www.learner.org/courses/worldhistory/unit_overview_3.html.

2. Litters: Moving Those Who Cannot Walk

Boyle, Laura. "Sedan Chairs." *World of Jane Austen* (June 20, 2011): http://www.janeausten.co.uk/sedan-chairs/.

"Huangshan Travel Tips." *China Hightlights* (January 26, 2015): http://www.chinahighlights.com/huangshan/travel-tips.htm.

"Litter." *Encyclopaedia Britannica*: http://www.britannica.com/EBchecked/topic/343924/litter.

"Sedan Chairs: An Efficient Mode of Transportation in Georgian London & Bath." *Jane Austen's World* (November 2, 2008): http://janeaustensworld.wordpress.com/2008/11/02/sedan-chairs-an-efficient-mode-of-transportation-in-georgian-london/.

Sedan Chair Charities Fund (2012–2015): http://www.sedanchairace.org/.

Additional Resources

"City of Bath." World Heritage Convention. UNESCO: http://whc.unesco.org/en/list/428.

"How Do I Build a Palanquin/Roman Litter?" Home Depot: http://community.homedepot.com/howto/DiscussionDetail/How-do-I-build-a-palanquin-roman-litter-9065000000007S3.

"Litter (Vehicle)." Wikipedia: http://en.wikipedia.org/wiki/Litter_(vehicle).

3. Dugouts: The First Boats

Africa's Oldest Known Boat: https://web.archive.org/web/20141030153747/http://wysinger.homestead.com/canoe.html.

Barklay-Kerr, Hoturoa. "Waka—Canoes." *Te Ara: The Encyclopedia of New Zealand* (December 13, 2013): http://www.teara.govt.nz/en/waka-canoes.

"Meet the Sea Gypsies." Burma Boating – Cruises in Myanmar (2014): http://www.burmaboating.com/cruising-area/.

Merfeld, Alison. "Facts on the Dugout Canoe." *eHow*: http://www.ehow.com/info_8637750_dugout-canoe.html.

Volmar, Mike. "The Dugout Canoe Project." Fruitlands Museum: https://web.archive.org/web/20150426031356/http://www.fruitlands.org/media/Dugout_Canoe_Article.pdf.

Additional Resources

"Dugout Canoe." Cherokee Heritage Center: http://www.cherokeeheritage.org/?page_id=241.

Ghose, Tia. "First Polynesians Arrived in Tonga 2,800 Years Ago." LiveScience (Nov 7, 2012): http://www.livescience.com/24614-first-polynesians-settlement.html.

Gillis, Bob. "How to Make a Dugout Canoe." Primitive Ways (2013): http://www.primitiveways.com/dugout-canoe.html.

"The Ocean Is Our Universe." [Moken Sea Gypsies.] Survival International: http://www.survivalinternational.org/galleries/moken-sea-gypsies.

"Polynesian History and Origin." Wayfinders: A Pacific Odyssey. WGBH: http://www.pbs.org/wayfinders/polynesian6.html.

4. Reed Rafts and Boats: As Old as Dugouts?

Allen, J.M. "History of Reed Ships": http://www.atlantisbolivia.org/areedboathistory.htm.

Dollinger, André. "Ancient Egyptian raw materials: Papyrus": http://www.reshafim.org.il/ad/egypt/trades/papyrus.htm.

———. "Ancient Egyptian Ships and Boats: The Archaeological Evidence": http://www.reshafim.org.il/ad/egypt/timelines/topics/navigation.htm.

Lawler, Andrew. "Report of Oldest Boat Hints at Early Trade Routes." *Science* 7 (June 2002): Vol. 296, no. 5574, pp. 1791–1792. DOI: 10.1126/science.296.5574.1791.

"Marine Technology": http://www.billbrouard.com/boat.htm.

Ryan, Donald P. "The Ra Expeditions Revisited": http://community.plu.edu/~ryandp/RAX.html.

"Thor Heyerdahl Expeditions and Archaeology of the Pacific Peoples": http://www.greatdreams.com/thor.htm.

Wikipedia. "Reed Boats": http://en.wikipedia.org/wiki/Reed_boat.

Additional Resources

AkaMya Talks: Traditional Tule Boat Making. [5½-minute video]: https://www.youtube.com/watch?v=L6ncwbKLEng.

Neuman, William, and Andrea Zarate. "Town's Floating Symbol Fading into the Sunset." *New York Times* (August 7, 2014): http://www.nytimes.com/2014/08/08/world/americas/towns-floating-symbol-fading-into-the-sunset.html.

"Thor Heyerdahl." *Wikipedia*: http://en.wikipedia.org/wiki/Thor_Heyerdahl.

Vranich, Alexei, Paul Harmon, and Chris Knutson. "Reed Boats and Experimental Archaeology on Lake Titicaca." *Expedition* 47, 2 (2005): pp. 20–27: http://www.penn.museum/documents/publications/expedition/pdfs/47-2/reed%20boats.pdf.

5. Planked Wooden Boats: The Shipbuilder's Art

"Galley," *Encyclopaedia Britannica*: http://www.britannica.com/EBchecked/topic/224325/galley.

Heyerdahl, Thor. *Early Man and the Ocean: A Search for the Beginnings of Navigation and Seaborne Civilizations*. New York: Doubleday & Co., Inc., 1979.

"History of Arms and Armour: Greek Fire, 674." HistoryWorld: http://www.historyworld.net/wrldhis/PlainTextHistories.asp?gtrack=pthc&ParagraphID=dlw#dlw.

Hornell, James. *Water Transport: Origins and Early Evolution*. Revised ed.; introduction by Basil Greenhill. [Newton Abbott, U.K.]: David & Charles, 1970.

Lienhard, John H. "The Last Galleys." *Engines of Our Ingenuity* 303 (1988–1997): http://www.uh.edu/engines/epi303.htm.

Paine, Lincoln P. *The Sea and Civilization: A Maritime History of the World*. New York, A. Knopf, 2013.

Additional Resources

"Actium (31 BC)." Livius.org (2003–2014): http://www.livius.org/battle/actium-31-bce/.

"Ancient Mariners of the Mediterranean: Exploring Shipwrecks to Unlock the Secrets of Ancient Civilizations." Video by National Geographic Education: http://education.nationalgeographic.com/education/media/ancient-mariners-mediterranean/?ar_a=1.

"Animated Map: Battle of Trafalgar." BBC History (2014): http://www.bbc.co.uk/history/interactive/animations/trafalgar/index_embed.shtml.

Biesty, Stephen, illustrator. "Building a Viking Cargo Ship": http://stephenbiesty.co.uk/galleries_Atmospheric_Cutaways_BuildingVikingCargoShip.html.

———, illustrator. "Columbus's Caravel": http://stephenbiesty.co.uk/galleries_Atmospheric_Cutaways_EverestExpedition.html.

———, illustrator. "Spanish Galleon": http://stephenbiesty.co.uk/galleries_cross_sections_galleon.html.

"Family Boatbuilding." *WoodenBoat* (July 21, 2011): http://www.woodenboat.com/family-boatbuilding.

Trueman, C.N. "The Spanish Armada." The History Learning Site (March 17, 2015): http://www.historylearningsite.co.uk/spanish_armada.htm.

Whitewright, Julian. "Roman Mediterranean Shipping." Archaeology of Portus: Exploring the Lost Harbour of Ancient Rome: http://moocs.southampton.ac.uk/portus/2014/06/03/roman-mediterranean-shipping/.

6. Sailors: Harnessing the Wind for at Least 6,000 Years

Australian National Maritime Museum, "Clipper Ships" [2007 exhibition]: http://www.anmm.gov.au/whats-on/exhibitions/archived-past-events/clipper-ships#About or Ahttps://web.archive.org/web/20140211213613/http://www.anmm.gov.au/site/page.cfm?u=1307.

"Clipper Ship." *Encyclopaedia Britannica*. http://www.britannica.com/EBchecked/topic/121871/clipper-ship.

Decker, Kris De. "Sailing at the Touch of a Button." *Low-Tech Magazine* (April 13, 2009): http://www.lowtechmagazine.com/2009/04/sailing-ships-large-crew-automated-control.html.

Fox, Stuart. "How Do Solar Sails Work?" *LiveScience* (May 17, 2010): http://www.livescience.com/32593-how-do-solar-sails-work-.html (accessed April 14, 2015).

NASA. "Windsurfing on a Wicked World." (May 1, 2012): http://www.nasa.gov/directorates/spacetech/home/feature_windsurfing.html.

Paine, Lincoln P. *The Sea and Civilization: A Maritime History of the World.* New York, A. Knopf, 2013.

Salem Maritime National Historic Site, "The Great Age of Duck," *Pickled Fish and Salted Provisions* 7, 4 (September 2005): http://www.nps.gov/sama/historyculture/upload/Vol7no4duck.pdf.

Space.com Staff. "Japanese Spacecraft Deploys Solar Sail." Space.com (June 11, 2010): http://www.space.com/8584-japanese-spacecraft-deploys-solar-sail.html.

Wall, Robert. "Rolls-Royce Revives Age of Sail to Beat Fuel-Cost Surge: Freight." *Bloomberg.com* (July 10, 2013): http://www.bloomberg.com/news/2013-07-10/rolls-royce-revives-age-of-sail-to-beat-fuel-cost-surge-freight.html.

"What Are Sails Made Out Of?" Answers.com: http://www.answers.com/Q/What_are_sails_made_out_of.

Additional Resources

Kilpatrick, Kevin. *Cutty Sark: A Beautiful Tea Clipper from the 1800s.* Greenwich, London HD, 2012. [video clip] https://www.youtube.com/watch?v=qwYREGkcGRc.

"Life at Sea in the Age of Sail." Royal Museums Greenwich. http://www.rmg.co.uk/explore/sea-and-ships/facts/ships-and-seafarers/life-at-sea-in-the-age-of-sail.

Pickled Fish and Salted Provisions, Salem Maritime National Historic Site, U.S. National Park Service [an occasional newsletter presenting "research on topics related to Salem Maritime, written in a fun and easy to read style"]: http://www.nps.gov/sama/learn/historyculture/pickledfish.htm.

Witek, Edward J. "The Ropewalks of Boston." Dr. Benjamin Church [blog] (September 30, 2011): http://drbenjaminchurchjr.blogspot.com/2011/09/ropewalks-of-boston.html.

Wolfe, Joe. "The Physics of Sailing." (2002): http://www.phys.unsw.edu.au/~jw/sailing.html.

7. Skis: Speed on the Slopes

Doyle, Mike. "Skiing History: The History of Downhill and Cross Country Skiing." About Sports, About.com: http://skiing.about.com/od/downhillskiing/a/skiinghistory.htm.

Fédération International de Ski. "History of Snow Sports." http://www.fis-ski.com/inside-fis/about/fis-history/snowsports/.

"A History of Skis." *National Geographic Magazine* (December 2013): http://ngm.nationalgeographic.com/2013/12/first-skiers/ski-history-interactive.

Lund, Morten, and Seth Masia, compilers. "A Short History of Skis." International Skiing History Association: https://skiinghistory.org/history/short-history-skis-0.

"Skiing History." Holmenkollen Ski Museum, Oslo, Norway: https://web.archive.org/web/20150225150441/http://www.holmenkollen.com/eng/The-Ski-Museum/Skiing-history.

Sood, Suemedha. "Where Did Skiing Come From?" BBC Travel (December 22, 2010): http://www.bbc.com/travel/blog/20101221-travelwise-where-did-skiing-come-from.

Additional Resources

Dayton, Kelsey. "High Art: A Ski Hunting Heritage." *Planet Jackson Hole* (January 21, 2014): http://planetjh.com/2014/01/21/high-art-a-ski-hunting-heritage/.

Piestrup, Zeke. "Guns, Blood & Skis: The History of Ski Warfare." The Ski Channel (August 7, 2009): http://www.theskichannel.com/news/20090807/guns-blood-skis-the-history-of-ski-warfare/.

"Polar Saga Part One." *National Geographic* (January 2009): http://ngm.nationalgeographic.com/2009/01/nansen/sides-text.

"Snowshoe Thompson." Wikipedia: http://en.wikipedia.org/wiki/Snowshoe_Thompson.

8. Horses: To Ride Like the Wind

British Museum, Ancient Civilizations, Mesopotamia. "Animals": http://www.mesopotamia. co.uk/staff/resources/background/bg27/home.html.

Cohen, Jennie. "Horse Domestication Happened Across Eurasia, Study Shows." History.com (January 30, 2012): http://www.history.com/news/ horse-domestication-happened-across-eurasia-study-shows.

"Deadliest Warrior – War Elephant." Wikia: http://deadliestwarrior.wikia.com/wiki/Elephant (accessed April 29, 2015).

"Draft Animal." *Encyclopaedia Britannica*. http://www.britannica.com/EBchecked/topic/170716/ draft-animal.

Klappenbach, Laura. "Domestication of Horses: The Relationship Between Horses and Humans." About.com: http://animals.about.com/od/hoofedmammals/a/domesticationof. htm.

Knodell, Kevin. "Battle of the Dumbos: Elephant Warfare from Ancient Greece to the Vietnam War," War is Boring (website): https://medium.com/war-is-boring/battle-of- the-dumbos-elephant-warfare-from-ancient-greece-to-the-vietnam-war-ca62af225917 (accessed April 29, 2015).

Levine, Marsha A. "Domestication, Breed Diversification and Early History of the Horse." Havemeyer Foundation: http://research.vet.upenn.edu/HavemeyerEquineBehaviorLab- HomePage/ReferenceLibraryHavemeyerEquineBehaviorLab/HavemeyerWorkshops/ HorseBehaviorandWelfare1316June2002/HorseBehaviorandWelfare2/Domestication- BreedDiversificationandEarlyHis/tabid/3127/Default.aspx.

Outram, Alan. "Horse Domestication in the Botai Culture, Eneolithic Kazakhstan." University of Exeter, Archaeology: https://humanities.exeter.ac.uk/archaeology/research/projects/ title_84579_en.html.

"War Elephants." http://www.hellenicaworld.com/Greece/LX/en/WarElephant.html.

"Western Zhou Chariot Burial Pit." TravelChinaGuide.com: http://www.travelchinaguide. com/attraction/shaanxi/xian/westernzhou.htm.

Additional Resources

Draft Animal-Power Network: http://www.draftanimalpower.org/.

"Hannibal's War Elephants." Hannibal Videos. History.com: http://www.history.com/topics/ ancient-history/hannibal/videos/hannibals-war-elephants.

Hirst, K. Kris. "Camel Caravan: Traveling the Silk Road—A Photoessay." About Education, About.com: http://archaeology.about.com/od/ancientcivilizations/ss/traveling_the_silk_ road_2.htm.

"Howdah." Wikipedia: http://en.wikipedia.org/wiki/Howdah.

Mosig, Yozan. "Hannibal's Elephants: Myth and Reality." *The History Herald* (18 May 2013): http://www.thehistoryherald.com/Articles/Ancient-History-Civilisation/ Hannibal-and-the-Punic-Wars/hannibal-s-elephants-myth-and-reality.

Quammen, David. "Native American Horses: People of the Horse." *National Geographic* (March 2014): http://ngm.nationalgeographic.com/2014/03/horse-tribes/quammen-text.

"Spanish Discovery of the New World." eXplorations. Digital History: http://www.digitalhis- tory.uh.edu/active_learning/explorations/spain/spain_menu.cfm.

"The Story of...Horses." Variables, Guns Germs & Steel. PBS: http://www.pbs.org/gunsgerms- steel/variables/horses.html.

9. The Wheel: Keep on Rolling

"Frequently Asked Questions." Swan Llamatrek: http://www.llamatreksmontana.com/ frequently-asked-questions/.

Turk, Jus. "Mysterious Pile-Dwellers: A Revelation about Prehistoric People in the Ljubljansko Barje. Slovenia.SI (April 2010): http://www.slovenia.si/slovenia/history/earliest-traces/ mysterious-pile-dwellers-a-revelation-about-prehistoric-people-in-the-ljubljansko-barje/.

Additional Resources

Cheesman, Paul. "The Wheel in Ancient America." *BYU Studies* 9, 2 (1969): http://shields-research.org/Scriptures/BoM/Wheel.htm.

Gambino, Megan. "A Salute to the Wheel." Smithsonian.com (June 17, 2009): http://www.smithsonianmag.com/science-nature/a-salute-to-the-wheel-31805121/?no-ist.

"Indus Valley Wheeled Ram Toy [Object]," in Children and Youth in History, Item #403: https://chnm.gmu.edu/cyh/primary-sources/403.

"Kids' Blog! The Invention of the Wheel: How the Ancient Sumeri-ans Got Humanity Rollin'!": http://antiquitynow.org/2013/08/20/ kids-blog-the-invention-of-the-wheel-how-the-ancient-sumerians-got-humanity-rollin/.

Schwaller, John F. "Why Did the Aztecs Only Use Wheels for Toys and Not for Transport?" Question for the Month, Aztecs, Mexico-lore (February 2010): http://www.mexicolore.co.uk/aztecs/ask-experts/ why-did-the-aztecs-only-use-wheels-for-toys-and-not-for-transport.

10. Chariots: Shock and Awe on the Battlefield

"The Chariot – The First War Machine." Making History, BBC Radio 4: http://www.bbc.co.uk/radio4/history/making_history/makhist10_prog5c.shtml.

Dollinger, André. "The Chariot": http://www.reshafim.org.il/ad/egypt/timelines/topics/chariot.htm.

Dunn, Jimmy (writing as Troy Fox). "The Chariot in Egyptian Warfare." TourEgypt.Net: http://www.touregypt.net/featurestories/chariots.htm.

NOVA. *Building the Pharaoh's Chariot* (video), originally aired February 6, 2013, on PBS: http://www.pbs.org/wgbh/nova/ancient/pharaoh-chariot.html.

Wilford, John Noble. "Remaking the Wheel: Evolution of the Chariot." *The New York Times Archive* (February 22, 1994): http://www.nytimes.com/1994/02/22/science/remaking-the-wheel-evolution-of-the-chariot.html.

Additional Resources

"Chariot Races." The Roman Empire in the 1ˢᵗ Century. PBS: http://www.pbs.org/empires/ romans/empire/chariot.html.

McManus, Barbara F. "The Circus: Roman Chariot Racing": http://www.vroma.org/~bmcmanus/circus.html.

Plubins, Rodrigo Quijada. "Chariot." Ancient History Encyclopedia (March 6, 2013): http://www.ancient.eu/chariot/.

11. Aerial Lifts: In Use More than 2,000 Years

Colorado School of Mines, Arthur Lakes Library. "About Ropeways." [2006]: http://inside.mines.edu/LIB-Ropeway-About; also found at http://web.archive.org/ web/20060904183704/http://www.mines.edu/library/ropeway/about_ropeways.html.

Decker, Kris De. "Aerial Ropeways: Automatic Cargo Transport for a Bargain." Shameez Joubert, ed. *Low-Tech Magazine* (January 2011): http://www.lowtechmagazine.com/2011/01/ aerial-ropeways-automatic-cargo-transport.html.

Engber, Daniel. "Who Made That Ski Lift?" *New York Times Magazine* (February 21, 2014): http://www.nytimes.com/2014/02/23/magazine/who-made-that-ski-lift.html?_r=0.

Sood, Suemedha. "Where Did Skiing Come From?" (December 22, 2010) BBC.com: http://www.bbc.com/travel/blog/20101221-travelwise-where-did-skiing-come-from.

Yarvin, Brian. "A Brief History of Ski Lifts and Cable Cars." *The Bend Bulletin* (February 1, 2012; updated November 19, 2013): http://www.bendbulletin.com/news/1351807-151/a-brief-history-of-ski-lifts-and-cable.

Additional Resources

"Aerial Ropeways." Open Source Ecology Wiki: http://opensourceecology.org/wiki/Aerial_Ropeways.

"Amazing Aerial Tramways of the World." Kuriositas (August 12, 2014): http://www.kuriositas.com/2011/02/amazing-aerial-tramways-of-world.html.

"Gravity Ropeways: The Clever Use of Gravitational Force Can Help People Get Their Produce to Market." Practical Action: http://practicalaction.org/gravity-ropeways-8.

12. Camels: Ships of the Desert

Australia.gov.au. *Afghan Cameleers in Australia*, http://www.australia.gov.au/about-australia/australian-story/afghan-cameleers (accessed May 27, 2015).

Bernstein, William J. *A Splendid Exchange: How Trade Shaped the World*. New York: Atlantic Monthly Press, 2008, p. 75.

ChinaVista.com. Camel Trains in the Desert: http://www.chinavista.com/experience/camel/camel.html (accessed May 12, 2015).

Encyclopaedia Britannica. *Caravan: Desert Transport*: http://www.britannica.com/EBchecked/topic/94606/caravan (accessed May 26, 2015).

Gish, Melissa. *Living Wild: Camels*. Mankato, MN, Creative Education, 2013, p. 32.

Pilkington, John. "Dying trade of the Sahara camel train," BBC News, Mali, last updated: Saturday, 21 October 2006, 11:26 GMT 12:26 U.K.: http://news.bbc.co.uk/2/hi/programmes/from_our_own_correspondent/6070400.stm (accessed May 22, 2015).

"Principal Deserts of the World." FactMonster.com: http://www.factmonster.com/ipka/a0778851.html (accessed May 26, 2015).

Rainier, Chris. "In Sahara, Salt-Hauling Camel Trains Struggle On." *National Geographic News*, May 28, 2003: http://news.nationalgeographic.com/news/2003/05/0528_030528_salt-caravan.html.

Thurbron, Colin. *Shadow of the Silk Road*. New York, HarperCollins Publishers, 2007.

Additional Resources

BBC News. "A Modern Camel Train in Pictures": http://news.bbc.co.uk/2/shared/spl/hi/pop_ups/06/africa_camel_train/html/1.stm.

CamelPhotos.com, the Camel hub of the Web: http://www.camelphotos.com/index.html.

National Geographic Sahara Photo Gallery – Photos by Chris Rainier that accompany his article on the salt caravans: http://news.nationalgeographic.com/news/2003/05/photogalleries/salt/.

13. Dogsleds: Driving Through the Snow

Bowers, Don. (Original document.) "The World Was Changing." History. Iditarod.com (edited 2012): http://iditarod.com/about/history/.

Klein, Christopher. "The Sled Dog Relay That Inspired the Iditarod." History.com (March 10, 2014): http://www.history.com/news/the-sled-dog-relay-that-inspired-the-iditarod.

Swan, Thom ("Swanny"). "'Marche': Sledge Dogs in the North West Fur Trade." *Stardancer Historical Freight Dogs*. Two Rivers, Alaska: http://www.tworiversak.com/sleddoghx1.htm.

Additional Resources

"A Brief History of the Sled" [photo gallery]. *Popular Mechanics*: http://www.popularmechanics. com/adventure/sports/g1377/a-brief-history-of-the-sled/.

"Diphtheria." U.S. Centers for Disease Control and Prevention: http://www.cdc.gov/diphtheria/index.html.

"The Early Years" [manufacturing diphtheria antitoxin—with the help of horses]. Wadsworth Center, New York Department of Health: http://www.wadsworth.org/about/history.

"Sled." Wikipedia: http://en.wikipedia.org/wiki/Sled.

14. Chinese Treasure Ships: Supertankers of the Ancient World

Hadingham, Evan. "Ancient Chinese Explorers." NOVA (January 16, 2001): http://www.pbs. org/wgbh/nova/ancient/ancient-chinese-explorers.html.

"Marco Polo, Explorer, Journalist (c. 1254–1324)." Biography.com (2015): http://www.biography.com/people/marco-polo-9443861.

Menzies, Gavin. *1421: The Year China Discovered America.* New York: HarperCollins, 2002.

Paine, Lincoln P. *The Sea and Civilization: A Maritime History of the World.* New York: A. Knopf, 2013.

Szczepanski, Kallie. "Zheng He's Treasure Ships." About Education, About.com: http://asianhistory.about.com/od/china/p/Zheng-Hes-Treasure-Ships.htm.

Additional Resources

Biesty, Stephen, illustrator. "Chinese Treasure Ship": http://stephenbiesty.co.uk/galleries_ Atmospheric_Cutaways_ChineseTreasureShip.html.

"Columbus's Ship vs. Zheng He's Treasure Ship" [video, comparing scale models]: https:// www.youtube.com/watch?v=FI1AmTa-bV0.

Furnish, Timothy. "Is Gavin Menzies Right or Wrong?" History News Network (George Mason University; March 13, 2003): http://historynewsnetwork.org/article/1308.

Hvistendahl, Mara. "Rebuilding a Treasure Ship." Archaeology Archive. *Abstracts* 61, 2 (March 2008): http://archive.archaeology.org/0803/abstracts/zhenghe.html.

Johnson, Ian. "Grand Canal: China's Ancient Lifeline." *National Geographic* (May 2013): http:// ngm.nationalgeographic.com/2013/05/chinas-grand-canal/johnson-text.

Menzies, Gavin. "1421." http://www.gavinmenzies.net/china/book-1421/.

"Voyages of Zheng He, 1405–1433" [map]. *National Geographic* (May 2007): http://ngm.nationalgeographic.com/ngm/0507/feature2/map.html.

15. Wheelchairs: Mobility for All

Bellis, Mary. "History of the Wheelchair." About Money, About.com: http://inventors.about. com/od/wstartinventions/a/wheelchair.htm.

"History of the Wheelchair." Mobility Scooters Otago. http://www.mobilityscooters.co.nz/ history/wheelchairs.

"History of Wheelchairs." Wheelchair Information (2013): http://www.wheelchair-information.com/history-of-wheelchairs.html.

Woods, Brian. "History of the Wheelchair." *Encyclopaedia Britannica* (February 6, 2014): http:// www.britannica.com/EBchecked/topic/1971423/history-of-the-wheelchair.

Additional Resources

Rosenhek, Jackie. "Before Wheelchairs." *Doctor's Review* (February 2007): http://www.doctorsreview.com/history/feb07-history_medicine/.

U.S. Department of Justice. *2010 ADA Standards for Accessible Design* (September 15, 2010): http:// www.ada.gov/regs2010/2010ADAStandards/2010ADAstandards.htm.

Wheelchair Basketball. http://www.paralympic.org/wheelchair-basketball.

Williamson, Bess. "The People's Sidewalks." *BOOM: A Journal of California* 2, 1 (Spring 2012): http://www.boomcalifornia.com/2012/06/the-peoples-sidewalks/.

16. Horse-Drawn Carriages: Traveling in Style

"Carriages: 17th Century." *History of Transport and Travel.* HistoryWorld.net: http://www.history-world.net/wrldhis/PlainTextHistories.asp?ParagraphID=kwq#1973.

Huggett, Frank E. *Carriages at Eight: Horse-drawn Society in Victorian and Edwardian Times.* New York: Charles Scribner's Sons, 1979.

Gay, Mara. "Carriage Horses, Now Controversial, Have a Long City History." Metropo-lis. *Wall Street Journal* (March 24, 2014): http://blogs.wsj.com/metropolis/2014/03/24/carriage-horses-now-controversial-have-a-long-city-history/.

Abdulaziz, Zainab. "History on Wheels: 5 Things to Know about the Queen's New Carriage. Today.com (June 4, 2014): http://www.today.com/news/history-wheels-5-things-know-about-queens-new-carriage-2D79756846.

Additional Resources

Robbins, Liz. "Who Speaks for the Carriage Horses?" [NYC Mayor De Blasio promises to end carriage drives around Central Park.] *New York Times* (January 17, 2014): http://www.nytimes.com/2014/01/19/nyregion/who-speaks-for-the-horses-in-battle-over-carriages.html.

Taylor, Kate. "After Carriage Horse Dies, Mayor Rejects Call for Ban." [NYC Mayor Bloomberg defends carriages around Central Park.] *New York Times* (October 26, 2011): http://www.nytimes.com/2011/10/27/nyregion/bloomberg-rejects-calls-to-ban-horse-drawn-carriages.html.

Wein, Josh. "Learn about the History of Horse-Drawn Carriages at the Long Island Museum" [video, 5½ minutes]. MLITV: https://www.youtube.com/watch?v=M15r6j2Xww4.

17. Yachts: Boating for Pleasure

Anderson, L.V. "What Is a Yacht? And When Did Owning One Become a Symbol of Wealth?" Slate.com (December 13, 2012): http://www.slate.com/articles/life/luxury_explainer/2012/12/what_is_a_yacht_and_why_are_they_associated_with_luxury.html.

"History of the America's Cup Races." 12 Meter Charters: http://12metercharters.com/americas-cup-race-history.

"The History of Luxury Yachts." Luxury Yachts Charter: http://www.megayachtscharter.com/en/general-information/history-of-luxury-yachts/.

Wheeler, Kelly L. "The History of Yachting." EzineArticles.com (April 2, 2008): http://ezinearticles.com/?The-History-of-Yachting&id=1084031.

"'Yacht': A Little History." Stability Yachts: http://www.stabilityyachts.com/yacht.html.

Additional Resources

America's Cup. "History": http://www.americascup.com/en/history.html.

"Cleopatra's Barge." Everything2 (May 18, 2013): http://everything2.com/title/Cleopatra%2527s+barge.

"Nemi Ships." Wikipedia: http://en.wikipedia.org/wiki/Nemi_ships.

"The New York Yacht Club": http://www.nyyc.org/about.

Warnes, Kathy. "Roman Emperor Caligula and His Legendary Lake Nemi Ships." History? Because It's Here: http://historybecauseitshere.weebly.com/roman-emperor-caligula-and-his-legendary-lake-nemi-ships.html.

18. Taxicabs: Wheels for Hire

Bellis, Mary. "Hailing – History of the Taxi: The Taxi was Named after the Taximeter." About Money, About.com: http://inventors.about.com/od/tstartinventions/a/taxi.htm.

Crank, Cindy. "Transport and Carriages in the Victorian era (1837–1901)." Horses and History Throughout the Ages Blog (May 2, 2011): https://horsesandhistory.wordpress.com/2011/05/02/transport-and-carriages-in-the-victorian-era-1837-1901/.

Huggett, Frank E. *Carriages at Eight: Horse-drawn Society in Victorian and Edwardian Times.* New York: Charles Scribner's Sons, 1979.

"June 1654, An Ordinance for the Regulation of Hackney-Coachmen in London and the Places Adjacent." Acts and Ordinances of the Interregnum, 1642–1660. British History Online: http://www.british-history.ac.uk/report.aspx?compid=56562.

"Stage Coach and Postchaise: 17th–18th Century." *History of Transport and Travel.* HistoryWorld.net: http://www.historyworld.net/wrldhis/PlainTextHistories.asp?ParagraphID=kwq#1973.

"Taxi History." Taxi Dreams. PBS.org: https://web.archive.org/web/20150310001737/http://www.pbs.org/wnet/taxidreams/history/.

U.S. Dept. of Labor, Bureau of Labor Statistics. "Taxi Drivers and Chauffeurs." *Occupational Outlook Handbook* (January 8, 2014): http://www.bls.gov/ooh/transportation-and-material-moving/taxi-drivers-and-chauffeurs.htm.

Additional Resources

"CAB!" *All the Year Round* [a magazine edited by Charles Dickens] (February 25, 1860): pp. 414–416. Accessed at http://www.taxi-library.org/dickens.htm.

"Old London Street Scenes (1903)" [early film; traffic shown includes hansom cabs]: https://www.youtube.com/watch?v=DVQiEJW7RWg&feature=relmfu.

Rubinstein, Diana. "Uber, Lyft, and the End of Taxi History." *Capital* (October 30, 2014): http://www.capitalnewyork.com/article/city-hall/2014/10/8555191/uber-lyft-and-end-taxi-history.

"World of Taxis—Culture and History" [annotated links]. Taxi Library: http://www.taxi-library.org/culture.htm.

19. Submarines: To the Bottom of the Sea

"DARPA Plans to Develop 'Flying Submarine.'" *Naval Technology News* (July 8, 2010): http://www.naval-technology.com/news/news89904.html.

DeepFlight corporate website: http://www.deepflight.com/ (accessed May 12, 2015).

Goebel, Greg. "The Invention of the Submarine." *In the Public Domain.* https://web.archive.org/web/20140316190136/http://www.vectorsite.net/twsub1.html.

Marks, Paul. "From Sea to Sky: Submarines That Fly." NewScientist.com (July 5, 2010): http://www.newscientist.com/article/mg20727671.000-from-sea-to-sky-submarines-that-fly.html#.VId_8V50xjo.

McLaughlin, Brett. "Cornelius Drebbel, Inventor of the Submarine." Dutchsubmarines.com: http://www.dutchsubmarines.com/specials/special_drebbel.htm (accessed April 27, 2015).

Triton Subs corporate website: http://tritonsubs.com/# (accessed May 12, 2015).

Additional Resources

McNicoll, Arion. "High-School Teen Builds One-Man Submarine for $2,000." CNN (May 29, 2013): http://www.cnn.com/2013/05/29/tech/innovation/teenager-justin-beckerman-builds-working-submarine/.

Milkovsky, Brenda. "David Bushnell and His Revolutionary Submarine." Connecticut History: http://connecticuthistory.org/david-bushnell-and-his-revolutionary-submarine/.

"The Periscope" [how to make one]. Science Toymaker: http://www.sciencetoymaker.org/periscope/index.html.

"Submarine History." Archive. Submarine Force Museum. http://ussnautilus.org/blog/category/submarine-history/.

Whitman, Edward C. "The Submarine Technology of Jules Verne." *Undersea Warfare: The Official Magazine of the U.S. Submarine Force* (Winter 2004): http://www.navy.mil/navydata/cno/n87/usw/issue_21/verne.htm.

20. Stagecoaches: Three Centuries of Service

"Coaching History." The Regency Collection: http://homepages.ihug.co.nz/~awoodley/carriage/history.html.

Larson, Elizabeth. "The Concord Coach." Over-Land.com (1996–2001): http://www.over-land.com/ccoach.html.

"Stagecoach Travel." *Dictionary of American History*. 2003. Encyclopedia.com (March 25, 2015): http://www.encyclopedia.com/doc/1G2-3401804002.html.

Additional Resources

"Mary Fields: Ex-Slave...Working in a Convent or Managing a Mail Route." BlackCowboys.com: http://www.blackcowboys.com/maryfields.htm.

Noyes, Larry. "The Stagecoaches: Their Origin and Place in History." *Tombstone Times*: http://www.tombstonetimes.com/stories/stagecoaches.html.

"Stage Travel in Britain." Georgian Index (2005): http://www.georgianindex.net/horse_and_carriage/coaching.html.

"Stagecoach." Wikipedia.

Wells Fargo. "Stagecoach History." http://www.wellsfargohistory.com/stagecoach/.

21. Buses: Another of Pascal's Great Ideas

Alfred, Randy. "March 18, 1662: The Bus Starts Here...in Paris." Wired.com (March 18, 2008): http://archive.wired.com/science/discoveries/news/2008/03/dayintech_0318.

Bus Stuff: History and Information on Buses Around the World (January 2011): http://www.bus-stuff.co.uk/2011/01/brief-history-of-bus.html.

"Electric Buses: Rapid Charging in Vienna." Siemens.com: http://www.siemens.com/innovation/en/home/pictures-of-the-future/mobility-and-motors/electric-mobility-electric-buses.html.

Golson, Jordan. "A Giant Charger That Juices Up Electric Buses in Three Minutes." Wired.com (October 2, 2014): http://www.wired.com/2014/10/giant-charger-juices-electric-buses-three-minutes/.

Huggett, Frank E. *Carriages at Eight: Horse-drawn Society in Victorian and Edwardian Times*. New York: Charles Scribner's Sons, 1979.

"A Short History of Public Transport in Greater Manchester." Museum of Transport, Greater Manchester (May 2004): https://web.archive.org/web/20150707163229/http://www.gmts.co.uk/explore/history/history.html.

"Stanislas Baudry et les omnibus de Nantes en 1825." (Source: http://www.amtuir.org - AMTUIR - Musée des Transports Urbains.) http://frenchinfluence.over-blog.fr/article-stanislas-baudry-et-les-omnibus-de-nantes-en-1825-68083889.html.

Additional Resources

Benzkofer, Stephan, and Ron Grossman. "Flashback: A Look at Chicago Transportation, From the Horse-Drawn Omnibus to Elevated Trains." *Chicago Tribune* (October 27, 2013): http://articles.chicagotribune.com/2013-10-27/news/ct-per-flash-transport-1027-20131027_1_streetcars-first-trolley-chicago-transportation.

"Bus Boycott in Alabama." America's Story from America's Library: http://www.americaslibrary.gov/aa/king/aa_king_bus_1.html.

Dowling, Stephen. "Radically Rethinking the Bus System." BBC Future (March 28, 2013): http://www.bbc.com/future/story/20130327-new-bus-stop-for-flexible-travel.

Greenfield, Adam. "Buses Are the Future of Urban Transport. No, Really." *The Guardian* (August 27, 2014): http://www.theguardian.com/cities/2014/aug/27/buses-future-of-urban-transport-brt-bus-rapid-transit.

Montgomery Bus Boycott: http://www.montgomeryboycott.com/.

Rodrigue, Jean-Paul. "Omnibus, London, Late 19th Century." Geography of Transport Systems (1998–2015): https://people.hofstra.edu/geotrans/eng/ch6en/conc6en/omni.html.

22. Automobiles: "Get a Horse!"

"Automobiles." History.com: http://www.history.com/topics/automobiles.

Bellis, Mary. "Gottfried Daimler." About Money, About.com: http://inventors.about.com/od/dstartinventors/a/Gottlieb_Daimler.htm.

———. "History of the Automobile." About.com: http://inventors.about.com/library/weekly/aacarssteama.htm.

Bottorff, William W. "What Was the First Car? A Quick History of the Automobile for Young People": https://web.archive.org/web/20140701050734/http://www.ausbcomp.com/~bbott/cars/carhist.htm.

Constable, George, and Bob Somerville. *A Century of Innovation: Twenty Engineering Achievements That Transformed Our Lives* (Washington, DC: National Academy of Engineering, 2003): accessed at: http://www.greatachievements.org/?id=2965 (accessed May 12, 2015).

Crouch, Tom D. *Wings: A History of Aviation from Kites to the Space Age.* Washington, DC: Smithsonian National Air and Space Museum, 2003.

Dick, Ron, and Dan Patterson. *Aviation Century: The Early Years.* Erin, Ontario: Boston Mills Press, 2003.

Falkvinge, Rick. "The Red Flag Act of 1865." TorrentFreak (June 26, 2011): https://torrentfreak.com/the-red-flag-act-of-1865-110626/.

"First American Oil Well." American Oil & Gas Historical Society: http://aoghs.org/oil-amanac/american-oil-history/ (accessed May 12, 2015).

"How We Use Energy: Transportation." The National Academies, What You Need to Know About Energy: http://needtoknow.nas.edu/energy/energy-use/transportation/ (accessed May 12, 2015).

Wise, David Burgess. *The Illustrated History of Automobiles.* New York: A & W Publishers, 1980.

Additional Resources

"Ultimate Wheels." History.com: http://www.history.co.uk/shows/ultimate-wheels.

How Stuff Works: How Automobiles Work; contains links to segments on how various components, such as the engine, brakes, tires, transmission, etc., work: http://auto.howstuffworks.com/automobile.htm.

How Stuff Works: How Car Engines Work; one of the links from the page above. Includes a short video on the "Innovation of the Engine" that gives a brief history of the engine, from steam to internal combustion. The video also includes a short plug for Valvoline Oil, which made the video: http://auto.howstuffworks.com/engine.htm.

U.S. Department of Transportation. "Don't Blame the Future." Highway History: http://www.fhwa.dot.gov/infrastructure/future.cfm.

23. Hot Air Balloons: Up, Up, and Away

Clark, Liesl. "A Short History of Ballooning." NOVA (December 2, 1997): http://www.pbs.org/wgbh/nova/space/short-history-of-ballooning.html.

Golomb, Jason. "Nasca Lines – The Sacred Landscape." *National Geographic*: http://science. nationalgeographic.com/science/archaeology/nasca-lines/.

"The History of Hot Air Ballooning." Arizona Balloon (2014): http://www.arizonaballoonclub. org/history_of_ballooning.htm.

"History of Hot Air Ballooning." eBalloon.org: http://www.eballoon.org/history/history-of-ballooning.html.

"Nazca Line Theories": http://www.bibliotecapleyades.net/nazca/esp_lineas_nazca_2.htm.

Nott, Julian, and Jim Woodman. "The Extraordinary Nazca Prehistoric Balloon." JulianNott. com (2014): http://juliannott.com/nazca/.

Additional Resources

Biesty, Stephen, illustrator. "Stratospheric Balloon Gondola": http://stephenbiesty.co.uk/galleries_Atmospheric_Cutaways_StratoshereBalloonGondola.html.

"Make a Hot Air Balloon from a Plastic Bag and Some Birthday Candles" [history and culture of ballooning; embedded video demonstration]. Science Toymaker: http://www.sciencetoymaker.org/HotAirBalloon/.

"Make a Mini Hot Air Balloon" [experimenting with air density; step-by-step instructions]. Home Science Tools: http://www.hometrainingtools.com/a/make-a-hot-air-balloon.

24. Steamships: Ocean Travel on a Schedule

"Canal History." New York State Canal Corporation: http://www.canals.ny.gov/history/history.html.

"A History of Steamboats." U.S. Army Corps of Engineers: http://www.sam.usace.army.mil/Portals/46/docs/recreation/OP-CO/montgomery/pdfs/10thand11th/ahistoryofsteamboats.pdf.

"Industrial Revolution." History.com: http://www.history.com/topics/industrial-revolution.

Othfors, Daniel, and Henrik Ljungström. "The Great Ocean Liners": http://www.thegreatoceanliners.com/index2.html.

"Ships and Steam Power." Royal Museums Greenwich: http://www.rmg.co.uk/explore/sea-and-ships/facts/ships-and-seafarers/steam-power.

"Steamships." The Open Door Web Site: http://www.saburchill.com/history/chapters/IR/033.html.

Additional Resources

Bellis, Mary. "James Watt—Inventor of the Modern Steam Engine." About Inventors, About. com: http://inventors.about.com/od/wstartinventors/a/james_watt.htm.

Lefler, Patrick. "Incremental Innovation and the Evolution of the Sailing Ship." Innovation Excellence (January 23, 2011): http://www.innovationexcellence.com/blog/2011/01/23/incremental-innovation-and-the-evolution-of-the-sailing-ship/.

"Paddle Power." [Design and build a paddle boat, powered by a rubber band.] DesignSquad Nation: http://pbskids.org/designsquad/parentseducators/resources/paddle_power.html.

"Paddle Steamship Animation." BBC History in Depth: http://www.bbc.co.uk/history/interactive/animations/paddlesteamer/index_embed.shtml.

"Speeding Up the Trade: Clippers and Steamships." Historical Collections Exhibit, Baker Library, Harvard Business School: http://www.library.hbs.edu/hc/heard/clippers-steamships.html.

Steamship Historical Society of America: https://www.sshsa.org/.

The Steamship Authority: "Lifeline to the Islands" since 1960: https://www.steamshipauthority.com/about/history.

25. Steam Locomotives: All Aboard!

Frew, Tim. *Locomotives: From the Steam Locomotive to the Bullet Train.* New York: Mallard Press, 1990.

Kashatus, William C. "Legend of Lackawanna Railroad's Phoebe Snow lives on," Citizensvoice. com (October 23, 2011): http://citizensvoice.com/arts-living/legend-of-lackawanna-rail-road-s-phoebe-snow-lives-on-1.1220950 (accessed April 28, 2015).

"Steam Train Anniversary Begins." BBC News (February 21, 2004): http://news.bbc.co.uk/2/hi/uk_news/wales/3509961.stm.

Rolt, L.T.C. "Richard Trevithick." *Encyclopaedia Britannica* (May 18, 2014): http://www.britannica.com/EBchecked/topic/604469/Richard-Trevithick.

Additional Resources

Biesty, Stephen, illustrator. "Steam Train": http://stephenbiesty.co.uk/galleries_cross_sections_steamtrain.html.

"A Chinese Immigrant Recalls the Dangers of Railroad Work." American Social History Project: http://herb.ashp.cuny.edu/items/show/614.

"Introduction: Transcontinental Railroad." *American Experience.* PBS: http://www.pbs.org/wgbh/americanexperience/features/introduction/tcrr-intro/.

Woodford, Chris. "Steam Engines." Explain That Stuff! (May 2014): http://www.explainthatstuff.com/steamengines.html.

26. Amphibious Vehicles: One If by Land, Two If by Sea?

Berg, Phil. "Swimming with Cars: 9 Amphibious Vehicles." *Popular Mechanics*: http://www.popularmechanics.com/cars/news/pictures/swimming-with-cars-9-amphibious-vehicles#slide-1.

"A Brief History of Amphibious Vehicles." Unique Cars and Parts: http://www.uniquecarsandparts.com.au/history_amphibians.htm.

"A History of Amphibious Cars, from the Amphicar to the Gibbs Sport Quadski." *New York Daily News* (October 22, 2012): http://www.nydailynews.com/autos/unusual-amphibious-cars-article-1.1189217.

"Channel Crossing Biography." Dover Museum website: http://www.dovermuseum.co.uk/Information-Resources/HPAC-Biographical/Channel-Crossing-Biography.aspx (accessed April 28, 2015).

Kettering, Charles F. "Orukter Amphibolos: A Radio Talk." *Short Stories of Science and Invention, A Collection of Radio Talks* (1942–1945; reprinted March 1959). Available at Today in Science History: http://todayinsci.com/stories2/story013.htm.

Lubar, Steven. "Was This America's First Steamboat, Locomotive, and Car?" *American Heritage's Innovation & Technology* 21, 4 (Spring 2006): http://www.innovationgateway.org/content/was-america%E2%80%99s-first-steamboat-locomotive-and-car-0.

Primm, Arallyn. "By Land and Sea: The History of Amphibious Vehicles." *Mental Floss* (May 28, 2013): http://mentalfloss.com/article/50725/land-and-sea-history-amphibious-vehicles.

U.S. Army Transportation Museum: DUKW (Duck): http://web.archive.org/web/20080725043610/http:/www.transchool.eustis.army.mil/Museum/DUKW.htm.

Additional Resources

Annear, Steve. "Designs Show How Boston Could Adapt as Sea Levels Rise." *Boston Globe* (February 19, 2015): http://www.bostonglobe.com/metro/2015/02/19/designs-show-how-boston-could-adapt-sea-levels-rise/ozvHRPXn1gPBKovHrDwE6O/story.html#.

Foynes Flying Boat & Maritime Museum: http://www.flyingboatmuseum.com/.

"Henri Fabre: 1882–1984" [inventor of the first seaplane]. Early Aviators: http://www.early-aviators.com/efabre.htm.

Motavalli, Jim. "Cars That Swim: The Cool Story of Amphibious Vehicles in the Marketplace." Car Talk (October 18, 2012): http://www.cartalk.com/content/cars-swim-cool-story-amphibious-vehicles-marketplace.

27. Covered Wagons: On the Oregon Trail

"American Western Migration: Wagon Trains and Covered Wagons." HistoryBits, Exploring True History (2013): http://www.historybits.com/west-wagon-trains.htm.

Flora, Stephenie, compiler. "The Covered Wagon." The Oregon Territory and Its Pioneers: http://www.oregonpioneers.com/wagon.htm.

Hill, William E. "Prairie Schooner." Encyclopaedia Britannica (2013): http://www.britannica.com/EBchecked/topic/473853/prairie-schooner.

Stewart, George R. "The Prairie Schooner Got Them There." American Heritage Magazine 13, 2 (February 1962): http://www.americanheritage.com/content/prairie-schooner-got-them-there.

Additional Resources

Flora, Stephenie, compiler. The Oregon Territory and Its Pioneers: http://www.oregonpioneers.com/ortrail.htm.

Oregon Trail Part I: https://www.youtube.com/watch?feature=player_detailpage&v=3CdRDFhwiqE.

Oregon Trail Part II: https://www.youtube.com/watch?feature=player_detailpage&v=zb2l3N47ttk.

The American West 03 – Wagon Trails to the West (1849), from Timelines.tv: https://www.youtube.com/watch?feature=player_detailpage&v=QsqpFuI6aQQ. Includes narration of diary entries from a young girl making the trip.

28. Ocean Liners: Huddled Masses in Steerage

Collins, Maria C. "Humanities 360: The Golden Age of Ocean Liners, 1900–1914." (March 11, 2013): https://web.archive.org/web/20140630172225/http://www.humanities360.com/index.php/the-golden-age-of-ocean-liners-1900-1914-1975/.

"Encyclopedia Titanica." Wikipedia: http://en.wikipedia.org/wiki/Encyclopedia_Titanica.

"Liners to America." On the Water, Exhibit 5. Smithsonian Institution: http://amhistory.si.edu/onthewater/exhibition/5_2.html.

"Ocean Liner." Wikipedia: http://en.wikipedia.org/wiki/Ocean_liner.

Othfors, Daniel, and Henrik Ljungström. "The Great Ocean Liners": http://www.thegreatoceanliners.com/index2.html.

Additional Resources

Ballard, Robert. "Lusitania." Excerpt from The Lost Liners. PBS Online: http://www.pbs.org/lostliners/lusitania.html.

Biesty, Stephen, illustrator. Queen Mary: http://stephenbiesty.co.uk/galleries_cross_sections_liner.html.

Encyclopedia Titanica: http://www.encyclopedia-titanica.org/.

"The Lusitania." World War I, History Learning Site: http://www.historylearningsite.co.uk/lusitania.htm.

"The Sinking of the Lusitania, 1915." EyeWitness to History (2000): http://www.eyewitnesstohistory.com/lusitania.htm.

29. Dirigibles

Crouch, Tom D. *Wings: A History of Aviation from Kites to the Space Age*. Washington, DC: Smithsonian National Air and Space Museum, 2003.

Grant, R.G. *Flight, The Complete History: 100 Years of Aviation*. New York: DK Publishing, 2002. 2007.

"Hindenburg Flight Schedule." Airships.net: A Dirigible and Zeppelin History Site: The Graf Zeppelin, Hindenburg, U.S. Navy Airships, and other Dirigibles: http://www.airships.net/hindenburg/flight-schedule.

"Jules Henri Giffard's Steam Airship." The Airship. TheHistoryForum.com: http://www.the-historyforum.com/airships/henri_giffard/.

Additional Resources

"Airships, Dirigibles, Zeppelins, & Blimps: What's the Difference?" Airships.net: A Dirigible and Zeppelin History Site: http://www.airships.net/dirigible.

Biesty, Stephen, illustrator. "Airship *Italia*": http://stephenbiesty.co.uk/galleries_Atmospheric_Cutaways_AirshipItalia.html.

Garber, Megan. "The Dead Dream of the Dirigible." *The Atlantic* (May 4, 2012): http://www.theatlantic.com/technology/archive/2012/05/the-dead-dream-of-the-dirigible/256758/.

Jones, Bryony. "Dawn of the Dirigibles: The New Age of the Airship?" CNN (June 21, 2013): http://edition.cnn.com/2013/06/21/business/new-age-of-the-airship/index.html.

30. Elevators: An Up and Down Business

"About Elevators." Otis Worldwide: http://www.otisworldwide.com/pdf/AboutElevators.pdf.

Bellis, Mary. "History of the Elevator: Elisha Otis." About Money, About.com: http://inventors.about.com/od/estartinventions/a/Elevator.htm.

"History of the Elevator." Elevators & Escalators, Mitsubishi Electric: http://www.mitsubishielectric.com/elevator/overview/elevators/history.html.

Newman, Ralph M. "Elevator Door Operators: Then and Now." *ElevatorWorld.com*: http://www.columbiaelevator.com/clientuploads/pdf/CE-Elevator-Door-Operators.pdf.

Additional Resources

"Elevator." Wikipedia: http://en.wikipedia.org/wiki/Elevator.

"Weight Changes in an Elevator – U.Va. Physics 109" [demonstration of Newton's Second Law]: https://www.youtube.com/watch?v=D-GzuZjawNI.

Woodford, Chris. "Elevators." ExplainThatStuff! (May 16, 2014): http://www.explainthatstuff.com/how-elevators-work.html.

31. Subways: Hurtling Through the Underworld

Bobrick, Benson. *Labyrinths of Iron: A History of the World's Subways*. New York: Newsweek Books, 1981.

Wilson, Tracy V. "How Subways Work." HowStuffWorks.com: http://science.howstuffworks.com/engineering/civil/subway.htm.

Kirby, Ed. "Inventor Frank Sprague." *Seldom Told Tales of Sharon*. Book Three. Sharon, CT: Sharon Historical Society: https://web.archive.org/web/20131219020028/http://www.sharonhist.org/inventor-frank-sprague.pdf.

Additional Resources

"About the MBTA: History" [Boston's pioneering 1897 subway system]: http://www.mbta.com/about_the_mbta/history/?id=962.

Biesty, Stephen, illustrator. "Subway Station": http://stephenbiesty.co.uk/galleries_cross_sections_tubestation.html.

Shafto, Dan. "Underground Railroads: A Short History of Subterranean Transport in America." Infoplease.com: http://www.infoplease.com/spot/subway1.html.

32. Bicycles: Pedal Power

Bicycle Collection. Canada Science and Technology Museum: https://web.archive.org/web/20131206220339/http://www.sciencetech.technomuses.ca/english/collection/cycles1.cfm.

The Bike-sharing World Map. Bikeshare.com: http://bikeshare.com/map/.

Fiedler, David. "Bikes – An Illustrated History." About Sports, About.com: http://bicycling.about.com/od/thebikelife/ss/History.htm.

Hendricks, David. "The Possibility of Mobility: The Rise and Fall of the Bicycle in 19th Century America" (2010): http://xroads.virginia.edu/~ug02/hendrick/bikehome.html.

"A Quick History of Bicycles." Pedaling History Bicycle Museum: https://web.archive.org/web/20140408032711/http://pedalinghistory.com/PHhistory.html.

Thomas, Pauline Weston. "Rational Dress Reform: Fashion History – Mrs Bloomer." Fashion-Era.com: http://www.fashion-era.com/rational_dress.htm (accessed April 28, 2015).

Additional Resources

"Bicycle History—The Nineteenth Century." Under Construction (November 30, 2010): https://martinsj2.wordpress.com/2010/11/30/bicycle-history-the-19th-century/.

Bloomfield, Susanne George, et al. "Bicycles, Bloomers, and the New Woman." Elia Peattie: An Uncommon Writer, an Uncommon Woman (1862–1935): http://plainshumanities.unl.edu/peattie/index.html.

The Tour de France website: http://www.letour.com/us/.

33. Motorcycles: Bicycles on Steroids

Bellis, Mary. "History of the Motorcycle." About.com: http://inventors.about.com/od/mstartinventions/a/motorcycle.htm.

Harris, William. "How Motorcycles Work." HowStuffWorks.com. http://auto.howstuffworks.com/motorcycle6.htm.

Hough, Richard, and L.J.K. Setright. History of the World's Motorcycles. New York: Harper & Row Publishers, 1966.

Long, Tony. "Aug. 30, 1885: Daimler Gives World First 'True' Motorcycle." This Day in Tech, 19th century. Wired (August 30, 2011): http://www.wired.com/2011/08/0830daimler-first-true-motorcycle/.

Smedman, Lisa. From Boneshakers to Choppers: The Rip-Roaring History of Motorcycles. Toronto: Annick Press, Ltd., 2007.

Additional Resources

Cameron, Kevin. "Where Is the Motorcycle Going? Is the Future Grand for Two Wheels?" Cycle World (November 21, 2012): http://www.cycleworld.com/2012/11/21/where-is-the-motorcycle-going/.

"The History of Motorcycle Helmets." Eagle Leather Motorcycle Gear (September 12, 2013): http://www.eagleleather.com/The-History-of-Motorcycle-Helmets-586/.

James, Randy. "A Brief History of the Hell's Angels." Time (August 3, 2009): http://content.time.com/time/nation/article/0,8599,1914201,00.html.

"Motorcycles." [Ecological arguments in favor of the motorcycle.] Green Choices: http://www.greenchoices.org/green-living/transport/motorcycles.

34. Rickshaws: Human-Powered Vehicles

Boland, Rory. "Rickshaw: Its History." About Travel, About.com: http://gohongkong.about. com/od/historyandcultureofhk/a/Rickshaw-A-Brief-History.htm.

Geens, Emily. "The History of the Rickshaw – Exploitation or Tradition?" *New Histories* 2, Issue 2.6 (April 14, 2011): http://newhistories.group.shef.ac.uk/wordpress/wordpress/ the-history-of-the-rickshaw-exploitation-or-tradition/.

Ray, Saikat. "Fresh Hope for Hand-Pulled Rickshaws." *The Times of India* (December 23, 2011): http://timesofindia.indiatimes.com/city/kolkata/Fresh-hope-for-hand-pulled-rickshaws/ articleshow/11212875.cms.

Shales, Melissa. "Durban's Ricksha Boys – Durban's Spectacular Rickshaw Runners." About Travel, About.com: http://goafrica.about.com/od/durbankwazulunatal/a/durban-rickshaw. htm.

Simpson, Margaret. "Japanese Rickshaw." Inside the Collection, Powerhouse Museum (November 19, 2012): http://www.powerhousemuseum.com/insidethecollection/tag/ jinrikisha/.

Trillin, Calvin. "The Last Days of the Rickshaw." *National Geographic* (April 2008): http://ngm. nationalgeographic.com/2008/04/kolkata-rickshaws/calvin-trillin-text.

Additional Resources

Kolkata's Rickshaws (slideshow): Venture into the world of Kolkata's hand-pulled rickshaw drivers with photographer Ami Vitale. Ami's field notes (a Q&A). A link to her video of the rickshaw drivers is in the upper left corner. http://ngm.nationalgeographic. com/2008/04/kolkata-rickshaws/vitale-field-notes.html.

Pictures of Chinese rickshaw driver Chen Guanming on his trip to Rio: http://english.agri.gov. cn/news/dqnf/201410/t20141013_24084.htm.

Zulu Ricksha, 1892–2000, Power Carriages of the Mandlakazi Clan: History of the evolution of the Durban, South Africa, Zulu rickshaw drivers and their amazing costumes, with photographs. Modern photos at the bottom are in color. http://www.ezakwantu.com/Gallery%20Zulu%20Ricksha.htm.

35. Cruise Ships: Floating Cities

"Cruise Ship." Wikipedia: http://en.wikipedia.org/wiki/Cruise_ship.

Hooper, John. "Italian Cruise Ship Fends Off Pirates with Gunfire." *The Guardian* (April 26, 2009): http://www.theguardian.com/world/2009/apr/26/italian-cruise-ship-pirates.

Slaten, Douglas D., and Kiren Mitruka. "Cruise Ship Travel" [in Chapter 6, "Conveyance and Transportation Issues," of *CDC Health Information for International Travel 2014*, ed. Gary W. Brunette, et al., Centers for Disease Control and Prevention]: http://wwwnc.cdc.gov/travel/ yellowbook/2014/chapter-6-conveyance-and-transportation-issues/cruise-ship-travel.

Viking Cruises: www.vikingcruises.com.

Additional Resources

"Shipboard Employment: Life on Board." Norwegian Cruise Line: http://www.ncl.com/about/ careers/shipboard-employment/life-onboard.

Sorabella, Jean. "The Grand Tour." Heilbrunn Timeline of Art History, the Metropolitan Museum of Art: http://www.metmuseum.org/toah/hd/grtr/hd_grtr.htm.

Tergesen, Anne. "Schools at Sea: A Look at Some of the Best Bets in Educational Cruises for Summer and Fall." *Wall Street Journal* (March 27, 2010): http://www.wsj.com/articles/SB100 01424052748704381604575005132158258728.

36. Airplanes: The Sky's the Limit

Crouch, Tom D. *Wings: A History of Aviation from Kites to the Space Age*. Washington, DC: Smithsonian National Air and Space Museum, 2003.

Dick, Ron, and Dan Patterson. *Aviation Century: The Early Years*. Erin, Ontario: Boston Mills Press, 2003.

Pendergast, Curtis, and the editors of Time-Life Books. *The Epic of Flight: The First Aviators*. Revised. Alexandria, VA: Time-Life Books, 1986.

Upton, Chad. "Why do Airplanes Fly at High Altitudes?" Broken Secrets (blog), posted February 23, 2010: http://brokensecrets.com/2010/02/23/why-do-airplanes-fly-at-high-altitudes/ (accessed April 18, 2015).

Additional Resources

"Ask History: Who Really Invented the First Airplane?" Wright Brothers Videos. History.com: http://www.history.com/topics/inventions/wright-brothers/videos/ask-history-who-really-invented-the-airplane.

"Bessie Coleman (1892–1926)" [first African-American woman pilot]. Fly Girls. *American Experience*, PBS: http://www.pbs.org/wgbh/amex/flygirls/peopleevents/pandeAMEX02.html.

Mayerowitz, Scott. "The Future of Aviation: Airline Leaders' Predictions." *USA Today* (January 1, 2014): http://www.usatoday.com/story/travel/flights/2014/01/01/future-of-aviation-airline-ceos-predictions/4267643/.

Miller, Don. "Barnstormers Gave Many Area Residents Their First Flying Experience." Mirror of History (September 17, 2012): https://web.archive.org/web/20121102152317/http://mirrorofhistory.areavoices.com/tag/barnstormers/.

"Secret Paper Aeroplanes" [instructions for 9 models]. Paper Aeroplanes.com: http://www.paperaeroplanes.com/.

37. Snowmobiles: Crossing Winter Landscapes

Allan, Dixie. "Snowmobiles." About Tech, About.com: http://webclipart.about.com/od/season-sclipart/ss/Snowmobiles.htm.

Allyson, Jane. "The History of Snowmobiles." Sporting Life 360 (January 22, 2008): https://web.archive.org/web/20140706025723/http://www.sportinglife360.com/index.php/the-history-of-snowmobiles-56187/.

BRP.com: http://www.brp.com/en-us.

Campbell, Stephen. "History of Snowmobiling." Northern Timber Cruisers Snowmobile Club [Museum]: http://www.millinocket-maine.net/history-of-snowmobiling.htm.

"Environmental Impacts from Snowmobile Use." Winter Wildlands Alliance (May 2014): http://winterwildlands.org/wp-content/uploads/2014/05/Environmental-Impacts-from-Snowmobile-Use.pdf (accessed April 29, 2015).

International Snowmobile Manufacturers' Association, Snowmobiling Fact Book: https://web.archive.org/web/20150523210556/http://snowmobile.org/snowmobilefacts.asp.

Wisconsin Snowmobile Ed Course. "History of Snowmobiles": https://www.snowmobile-ed.com/wisconsin/studyGuide/History-of-Snowmobiles/50105101_700066571.

Wuerthner, George, ed. *The Environmental Consequences of Motorized Recreation* (San Francisco, CA: The Foundation for Deep Ecology, 2008) accessed at http://www.stopthrillcraft.org/kind_snowmobiles.htm (accessed April 29, 2015).

Additional Resources

Barabak, Mark Z. "Midterm Madness: The Politics of Snowmobiles and Chickens." *LA Times* (October 1, 2014): http://www.latimes.com/nation/politics/politicsnow/la-pn-midterm-election-politics-snowmobiles-chickens-20140930-story.html#page=1.

Barringer, Felicity. "Judge Voids New Rule Allowing Snowmobiles in Yellowstone." *New York Times* (December 17, 2003): http://www.nytimes.com/2003/12/17/politics/17PARK.html.

International Snowmobile Manufacturers' Association. "Snowmobiling Fact Book: Effects of Snowmobiling on..." [addresses Yellowstone National Park concerns]: https://web.archive.org/web/20150420035740/http://www.snowmobile.org/facts_ece.asp.

"Safe Riders!" Snowmobile Safety Awareness Program: http://www.saferiderssafetyawareness.org/.

38. Recreational Vehicles (RVs): Cruising on Land

"History of the Vardo (Gypsy Caravan)." Valley Stream Media, 1980–2014: http://gypsywaggons.co.uk/varhistory.htm.

"The History of the RV in Canada and the U.S." RV Hotline: http://www.rvhotlinecanada.com/rvhistory.asp.

Gothard, Julian. "The World's Ten Best Car Carrying Motorhomes." Examiner.com (August 26, 2012): http://www.examiner.com/article/the-world-s-ten-best-car-carrying-motorhomes.

Additional Resources

Adams, Jeff. "A Short History of Camping." ReserveAmerica.com: http://www.reserveamerica.com/outdoors/a-short-history-of-camping.htm.

"John Muir (1838–1914)" [on a 3-night camping trip with Theodore Roosevelt, he advocated national parks]. Ken Burns, *The National Parks*, WGBH: http://www.pbs.org/nationalparks/people/historical/muir/.

Klein, Christopher. "Ford and Edison's Excellent Camping Adventures." History in the Headlines (July 30, 2013): http://www.history.com/news/ford-and-edisons-excellent-camping-adventures.

Pappas, Stephanie. "Origin of the Romani People Pinned Down." *LiveScience* (December 6, 2012): http://www.livescience.com/25294-origin-romani-people.html.

39. Helicopters: Da Vinci's Dream

"Crane worker saved from Kingston fire in 'extremely unusual rescue.'" *CBC News* (last updated December 18, 2013): http://www.cbc.ca/news/canada/ottawa/crane-worker-saved-from-kingston-fire-in-extremely-unusual-rescue-1.2467711 (accessed May 5, 2015).

Crouch, Tom D. *Wings: A History of Aviation from Kites to the Space Age*. Washington, DC: Smithsonian National Air and Space Museum, 2003.

"Helicopter (Aerial Screw)." Leonardo da Vinci Inventions. InventHelp: www.da-vinci-inventions.com/aerial-screw.aspx.

Leishman, J. Gordon. *Principles of Helicopter Aerodynamics*. Cambridge aerospace series, 18. Cambridge, U.K.: Cambridge University Press, 2006. ISBN 978-0-521-85860-1. Web extract.

Wise, Jeff. "Finally! A Human-Powered Helicopter Wins the $250,000 Sikorsky Prize." *Popular Mechanics* (July 11, 2013): http://www.popularmechanics.com/flight/how-to/a9138/finally-a-human-powered-helicopter-wins-the-250000-sikorsky-prize-15682369/ (accessed May 5, 2015).

Withnall, Adam. "Leonardo Da Vinci's dream realised: Canadian inventors achieve world's first man-powered helicopter flight." (July 12, 2013), *The Independent*: http://www.independent.co.uk/news/world/americas/leonardo-da-vincis-dream-realised-canadian-inventors-achieve-worlds-first-manpowered-helicopter-flight-8706413.html (accessed May 5, 2015).

Young, Warren R., and the editors of Time-Life Books. *The Epic of Flight: The Helicopters*. Alexandria, VA: Time-Life Books, 1982.

Additional Resources

Biesty, Stephen, illustrator. "Rescue Helicopter": http://stephenbiesty.co.uk/galleries_cross_sections_helicopter.html.

"How Helicopters Work" [video]. HowStuffWorks.com: http://science.howstuffworks.com/transport/4728-how-helicopters-work-video.htm.

Tims, Dana. "Dramatic Helicopter Rescue of Five Hikers Stranded on Ocean Rocks near Depoe Bay Caught on Video." *Oregon Live, The Oregonian* (October 12, 2014): http://www.oregonlive.com/pacific-northwest-news/index.ssf/2014/10/dramatic_helicopter_of_five_hi.html.

40. Pedicabs: A Cycle Rickshaw by Any Other Name

Burchell, Helen. "Chen Guanming: The Olympic Chinese Rickshaw Rider's Farewell to England." BBC News England (August 23, 2013): http://www.bbc.com/news/uk-england-23792836.

"Chinese Farmer Travels from Canada to Brazil with Olympic Rickshaw." Ministry of Agriculture of the Peoples Republic of China web site, Xinhua News Agency (October 13, 2014): http://english.agri.gov.cn/news/dqnf/201410/t20141013_24084.htm.

Dixon, Phillip. "Pedicabs Endure a Rough Ride." *Bicycle Times Magazine* (December 29, 2009): http://bicycletimesmag.com/pedicabs-endure-a-rough-ride/.

Lao Long. "Rickshaw." Archived in Culture section, Newsfinder (June 2, 2003): http://www.newsfinder.org/site/more/rickshaw/.

Le Tran, Ann. "Rickshaws, Elvis, and Outside Lands: The History of Pedicabs." Esurance blog (July 24, 2012): http://blog.esurance.com/history-of-pedicabs/#.VFDPcl5oxjo.

Additional Resources

Central Park Pedicab Tours: http://www.centralparkpedicabs.com/.

"Pedicab Driver Charged Tourists $720 for 20-Minute Ride: Report." News 4 New York (July 19, 2013): http://www.nbcnewyork.com/news/local/Pedicab-Driver-Rip-Off-Charge-Couple-720-Dollars-Japanese-Tourist-Consumer-Affairs-216139711.html.

Schaller, Bruce. "Future of Pedicabs Remains Uncertain." *Gotham Gazette* (April 18, 2007): http://www.gothamgazette.com/index.php/transportation/3531-future-of-pedicabs-remains-uncertain.

41. Skateboards: Have Board, Will Travel

Behre, Robert. "Skateboarders Face Legal Roadblocks." *The Post and Courier*, Charleston, SC (April 30, 2011): http://www.postandcourier.com/article/20110430/PC1602/304309937.

Bradley, Daniel. "Skateboards: Fit & Types." LiveStrong.com (October 30, 2013): http://www.livestrong.com/article/339215-skateboards-fit-types/.

Cave, Steve. "A Brief History of Skateboarding." About Sports, About.com: http://skateboard.about.com/cs/boardscience/a/brief_history.htm.

"Go Skateboarding Day." International Association of Skateboard Companies: http://theiasc.org/go-skateboarding-day/.

Rompella, Natalie. *Famous Firsts: The Trendsetters, Groundbreakers & Risk-Takers Who Got America Moving!* Lobster Press, 2007: http://books.google.com/books?id=bc-_uWnoZWwC&pg=PA30#v=onepage&q&f=false.

Additional Resources

Exploratorium: Skateboard Science — Find out how this extreme sport is governed by the principles of momentum, gravity, friction, and centripetal force. This site looks at the physics behind skateboard tricks: http://www.exploratorium.edu/skateboarding/.

Owen, Tony. "The Evolution of Skateboarding—A History from Sidewalk Surfing to Super-stardom." *Skateboarding Magazine* (March 5, 2013): http://www.skateboardingmagazine.com/the-evolution-of-skateboarding-a-history-from-sidewalk-surfing-to-superstardom/.

42. Hovercraft: Riding a Cushion of Air

Herring, Sharolyn, and Christopher Fitzgerald. "History of the Hovercraft." World Hovercraft Organization and Neoteric Hovercraft, Inc.: http://www.neoterichovercraft.com/general_info/historyof.htm.

"History of Hovercraft." DiscoverHover: http://www.discoverhover.org/abouthovercraft/history.htm.

"Hovercraft History." BBC Home, Hampshire & Isle of Wight (March 28, 2008): http://www.bbc.co.uk/hampshire/content/articles/2008/03/27/history_hovercraft_feature.shtml.

Additional Resources

The Build-a-Hovercraft School Project. DiscoverHover: http://www.discoverhover.org/index.htm.

Build Hovercraft. DesignSquad Nation: http://pbskids.org/designsquad/build/hovercraft/.

HubPages, History of the Hovercraft, by Tom Payne – First part is the history of the hovercraft, but the second part (about halfway down the page) gives a good illustration and explanation of how a hovercraft works: http://tonypayne.hubpages.com/hub/hovercraft-history.

43. Electric and Hybrid Automobiles: So Nineteenth Century

Bellis, Mary. "History of Electric Vehicles – The Early Years, Electric Cars from 1830 to 1930." About.com: http://inventors.about.com/od/estartinventions/a/History-Of-Electric-Vehicles.htm.

Berman, Brad. "History of Hybrid Vehicles." Hybridcars.com (June 14, 2011): http://www.hybridcars.com/history-of-hybrid-vehicles/.

Keane, Angela Greiling. "US Will Require Too-Quiet Electric Cars to Make Noise." *Boston Globe* (January 8, 2013): http://www.bostonglobe.com/business/2013/01/08/government-require-electric-cars-make-noise/bGmucrhawg3ftkIi7HqdQO/story.html.

Lake, Matt. "A Tale of 2 Engines: How Hybrid Cars Tame Emissions." *New York Times*, Technology (November 8, 2001): http://www.nytimes.com/2001/11/08/technology/how-it-works-a-tale-of-2-engines-how-hybrid-cars-tame-emissions.html?scp=1&sq=hybrid%20Toyota%20Prius%201997%20Honda%20Insight%201999&st=cse.

Rogowsky, Mark. "6 Reasons Tesla's Battery Swapping Could Take It to a 'Better Place.'" *Forbes Magazine* June 21, 2013): http://www.forbes.com/sites/markrogowsky/2013/06/21/6-reasons-teslas-battery-swapping-could-take-it-to-a-better-place/.

Zehner, Ozzie. "Unclean at Any Speed: Electric Cars Don't Solve the Automobile's Environmental Problems." *IEEE Spectrum* (June 30, 2013): http://spectrum.ieee.org/energy/renewables/unclean-at-any-speed.

Additional Resources

"All About Electric Vehicles." Car Talk. http://www.cartalk.com/content/all-about-electric-vehicles.

Brain, Marshall. "How Electric Cars Work" [includes video, "Why Aren't There More Electric Cars?" suggesting conspiracy between automakers and oil industry]. How Stuff Works: http://auto.howstuffworks.com/electric-car.htm.

Cooper Hewitt. "Why Design Now? MIT CityCar" [5-minute video; engineless city automobile of the future]: https://www.youtube.com/watch?v=y77txJ64RBg.

"Green Travel Plans." ETA: https://www.eta.co.uk/environmental-info/green-travel-plans/.

Lacey, Stephen. "Would Solar Roadways Work? A Government Engineer Discusses the Controversial Technology." Greentech Solar (August 29, 2014): http://www.greentechmedia.com/articles/read/Department-of-Transportation-Official-Discusses-Solar-Roadways.

McLachlan, Justin. "How Intelligent Roads—Not Just Cars—Will Change Transportation." *Popular Science* (April 1, 2013): http://www.popsci.com/technology/article/2013-04/street-smarts.

44. All-Terrain Vehicles: A Ticket to the Back Country

Bansal, V., et al. "A 21-year History of All-Terrain Vehicle Injuries: Has Anything Changed?" *American Journal of Surgery* 195, 6 (June 2008) [abstract]: See comment in PubMed Commons belowhttp://www.ncbi.nlm.nih.gov/pubmed/18367134?report=abstract.

"The Environmental Impact of ATVs." Portland Natural Gas Transmission System: http://www.pngts.com/atv.html.

"Facts and History about ATV." ATV Quad News. ATV.info: http://www.atv.info/page.cfm?name=ATV%20Facts.

"The History of the ATV." Outdoor Hub: http://www.outdoorhub.com/stories/2012/08/01/the-history-of-the-atv/.

Additional Resources

Santos, Zeon. "The Amphibious All-Terrain Vehicle of the Future" [links to 14-minute video]. Neatorama (May 29, 2012): http://www.neatorama.com/2012/05/29/the-amphibious-all-terrain-vehicle-of-the-future/.

45. Manned Spacecraft: Out of This World

Crouch, Tom D. *Wings: A History of Aviation from Kites to the Space Age.* Washington, DC: Smithsonian National Air and Space Museum, 2003.

Grant, R.G. *Flight, The Complete History: 100 Years of Aviation.* New York: DK Publishing, 2002. 2007.

Braeunig, Robert A. "Did We Land on the Moon? A Debunking of the Moon Hoax Theory": http://braeunig.us/space/hoax.htm (accessed May 6, 2015).

Additional Resources

Anderson, Chris. "Elon Musk's Mission to Mars." *Wired* (October 21, 2012): http://www.wired.com/2012/10/ff-elon-musk-qa/all/.

Moore, Trent. "Bill Nye and Neil DeGrasse Tyson's New Spacecraft Set for First Test Flight." Blastr (January 26, 2015): http://www.blastr.com/2015-1-26/bill-nye-and-neil-degrasse-tysons-new-spacecraft-set-first-test-flight.

NASA. "Exploring Careers @ NASA: For Students": http://www.nasa.gov/audience/forstudents/careers-index.html.

NASA. "One Year Mission and Twins Study: Human Research": http://www.nasa.gov/content/twins-study/.

NASA Jet Propulsion Laboratory. "Spacecraft: Surface Operations: Rover" [a vehicle that doesn't take us along—or does it?]: http://mars.nasa.gov/mer/mission/spacecraft_surface_rover.html.

46. Jetpacks: Strap One On and Fly

Jetman: http://www.jetman.com/?page_id=24.

The Martin Jetpack: http://www.martinjetpack.com/.

Ricker, Thomas. "The Future Passed: Jetpack Edition." *The Verge* (November 3, 2011): http://www.theverge.com/2011/11/3/2504531/jetpack-history-future-passed.

Additional Resources
"Almanac: Rocket Belt." CBS (April 20, 2014): http://www.cbsnews.com/news/almanac-rocket-belt/.
Burnett, Dean. "Jetpacks: Here's Why You Don't Have One." *The Guardian* (September 23, 2014): http://www.theguardian.com/science/brain-flapping/2014/sep/23/jetpacks-science-scientists.
Egan, Matt. "Jetpacks are Real—and They're Awesome." CNN Money (November 5, 2014): http://money.cnn.com/2014/11/05/luxury/jetpack-travel-luxury-water-sports/.

47. Bullet Trains: High-Speed Rail

"China High-Speed Train (Bullet Train)." TravelChinaGuide.com: http://www.travelchinaguide.com/china-trains/high-speed/.
Frew, Tim. *Locomotives: From the Steam Locomotive to the Bullet Train*. New York: Mallard Press, 1990.
High-Speed Europe: A Sustainable Link Between Citizens. Luxembourg: Publications Office of the European Union, 2010: http://ec.europa.eu/transport/themes/infrastructure/studies/doc/2010_high_speed_rail_en.pdf.
Peterman, David Randall (coordinator), John Frittelli, and William J. Mallett. *The Development of High Speed Rail in the United States: Issues and Recent Events.* Congressional Research Service (December 20, 2013): http://fas.org/sgp/crs/misc/R42584.pdf.
"Shinkansen." Japan-guide.com (1996–2015): http://www.japan-guide.com/e/e2018.html.

Additional Resources
"The Fifteen Billion Pound Railway, Episode 3: Platforms and Plague Pits." BBC Documentary (2014): https://www.youtube.com/watch?v=DlziPDJXQbQ.
Kennedy, Bruce. "How Viable Is a U.S. High-Speed Rail System?" Moneywatch. CBS News (October 21, 2014): http://www.cbsnews.com/news/how-viable-is-a-u-s-high-speed-rail-system/.
Science – How Stuff Works – How Maglev Trains Work [includes a video of a train]: http://science.howstuffworks.com/transport/engines-equipment/maglev-train.htm.
U.S. Department of Transportation. "Adapting to Climate Change": http://www.dot.gov/mission/sustainability/adapting-climate-change.
U.S. Environmental Protection Agency. "Climate Impacts on Transportation": https://www3.epa.gov/climatechange/impacts/transportation.html.

48. Supersonic Airliners: Breaking the Sound Barrier

Crouch, Tom D. *Wings: A History of Aviation from Kites to the Space Age.* Washington, DC: Smithsonian National Air and Space Museum, 2003.
Grant, R.G. *Flight, The Complete History: 100 Years of Aviation.* New York: DK Publishing, 2002. 2007.

Additional Resources
"Air Travel's Impact on Climate Change." ETA: https://www.eta.co.uk/environmental-info/air-travels-impact-on-climate-change/.
"Celebrating Concorde." History and Heritage. British Airways: http://www.britishairways.com/en-us/information/about-ba/history-and-heritage/celebrating-concorde.
U.S. Environmental Protection Agency. "Climate Impacts on Transportation": https://www3.epa.gov/climatechange/impacts/transportation.html.

Westcott, Richard. "Could Concorde Ever Fly Again? No, Says British Airways." BBC News (October 24, 2013): http://www.bbc.com/news/business-24629451.

49. Space Shuttles: Into Space and Back Again

Crouch, Tom D. *Wings: A History of Aviation from Kites to the Space Age.* Washington, DC: Smithsonian National Air and Space Museum, 2003.
Grant, R.G. *Flight, The Complete History: 100 Years of Aviation.* New York: DK Publishing, 2002. 2007.

Additional Resources

Biesty, Stephen, illustrator. "Space Shuttle": http://stephenbiesty.co.uk/galleries_cross_sections_spaceshuttle.html.
Freudenrich, Craig. "How Space Shuttles Work." HowStuffWorks: http://science.howstuffworks.com/space-shuttle.htm.
NASA. "Challenger STS 51-L Accident" [NASA links]: http://www.history.nasa.gov/sts51l.html.
NASA. Featured Images: The Space Shuttle Era: http://www.nasa.gov/mission_pages/shuttle/main/index.html.
NASA. Space Shuttle Web archives: http://spaceflight.nasa.gov/home/index.html.
"30 Years of the Space Shuttle" [graphic timeline with links]. *New York Times* (July 21, 2011): http://www.nytimes.com/interactive/science/space/0705-shuttle-30-years.html?_r=0.

50. The Segway: Pedestrians on Wheels

Governors Highway Safety Association, Segway Laws: http://www.ghsa.org/html/stateinfo/laws/segway_laws.html.
Harris, Tom. "How Segways Work." Science, How Stuff Works: http://science.howstuffworks.com/engineering/civil/ginger.htm.
"Segway, Inc." Wikipedia: http://en.wikipedia.org/wiki/Segway_Inc.
Segway Inc. website (2015): http://www.segway.com/.
"Segway Polo." Wikipedia: http://en.wikipedia.org/wiki/Segway_polo.
Spielman, Fran. "City Puts Brake on Segway Tours." *Chicago Sun-Times* (May 28, 2014): https://web.archive.org/web/20151003211130/http://chicago.suntimes.com/?p=165263.

Additional Resources

Chin, Ryan. "Solving Transport Headaches in the Cities of 2050." BBC Future (June 18, 2013): http://www.bbc.com/future/story/20130617-moving-around-in-the-megacity.
Harris, Tom. "How Segways Work." How Stuff Works: http://science.howstuffworks.com/engineering/civil/ginger.htm.
Nizza, Mike. "Segways Grow Old Nerdily." *New York Times* (July 8, 2008): http://thelede.blogs.nytimes.com/2008/07/08/segways-grow-old-nerdily/.

Endnotes

Introduction

1. "Steve Fossett," Biography.com. http://www.biography.com/people/steve-fossett-9542064 (accessed January 7, 2016).

Chapter 1

1. "This Day in History, July 20, 1969, Armstrong walks on moon," History.com. http://www.history.com/this-day-in-history/armstrong-walks-on-moon (accessed October 7, 2014).
2. Guy Gugliotta, "The Great Human Migration: Why humans left their African homeland 80,000 years ago to colonize the world," *Smithsonian Magazine* (July 2008): http://www.smithsonianmag.com/history/the-great-human-migration-13561/?page=5 (accessed October 7, 2014).
3. Tom Connolly, "The World's Oldest Shoes," University of Oregon web site. http://pages.uoregon.edu/connolly/FRsandals.htm (accessed October 8, 2014).
4. Kate Ravilious, "World's Oldest Leather Shoe Found—Stunningly Preserved," *National Geographic News* (June 9, 2010): http://news.nationalgeographic.com/news/2010/06/100609-worlds-oldest-leather-shoe-armenia-science/ (accessed October 7, 2014).
5. Maggie Koerth-Baker, "First Shoes Worn 40,000 Years Ago," *LiveScience* (June 5, 2008): http://www.livescience.com/4964-shoes-worn-40-000-years.html (accessed October 7, 2014).
6. Ravilious.
7. Simon Quellen Field, "What Are Shoes Made From?" *200 Questions About Chemistry*. http://questions.sci-toys.com/node/65 (accessed October 7, 2014).
8. J.K Gillion, "Robert Barclay Allardice: The Celebrated Pedestrian." http://gillonj.tripod.com/thecelebratedpedestrian/ (accessed October 7, 2014).
9. Mike Rosenbaum, "An Illustrated History of Race Walking," About Sports. http://trackandfield.about.com/od/distanceevents/ss/illusracewalk.htm (accessed October 7, 2014).
10. This Day in History, July 20, 1969.

Chapter 2

1. "Litter," *Encyclopaedia Britannica*. http://www.britannica.com/EBchecked/topic/343924/litter (accessed October 7, 2014).
2, Laura Boyle, "Sedan Chairs," *The Jane Austen Centre* (blog), June 20, 2011. http://www.janeausten.co.uk/sedan-chairs/ (accessed October 8, 2014).
3. "Sedan Chairs: An Efficient Mode of Transportation in Georgian London & Bath," *Jane Austen's World* (blog), September 2, 2008. http://janeaustensworld.wordpress.com/2008/11/02/sedan-chairs-an-efficient-mode-of-transportation-in-georgian-london/.
4. "Sedan Chairs."
5. "Huangshan Travel Tips," *China Highlights*. http://www.chinahighlights.com/huangshan/travel-tips.htm (accessed October 8, 2014).
6. Sedan Chair Race Charities Fund, http://www.sedanchairace.org/ (accessed October 8, 2014).

Chapter 3

1. Alison Merfeld, "Facts on the Dugout Canoe," eHow.com. http://www.ehow.com/info_8637750_dugout-canoe.html (accessed October 7, 2014).

2. Mike Volmar, "The Dugout Canoe Project," Fruitlands.org. https://web.archive.org/web/20150426031356/http://www.fruitlands.org/media/Dugout_Canoe_Article.pdf.

3. "Africa's Oldest Known Boat," http://wysinger.homestead.com/canoe.html (accessed October 7, 2014).

4. "Waka – canoes," *Teara, The Encyclopedia of New Zealand*. http://www.teara.govt.nz/en/waka-canoes (accessed October 7, 2014).

5. "Our Sailing Area, the Mergui Archipelago," *Burma Boating – Cruises in Myanmar*, http://www.burmaboating.com/cruising-area/ (accessed October 8, 2014).

Chapter 4

1. André Dollinger, "Papyrus," *An introduction to the history and culture of Pharaonic Egypt*, http://www.reshafim.org.il/ad/egypt/trades/papyrus.htm (accessed June 28, 2014).

2. Donald P. Ryan, "The Ra Expeditions Revisited," http://community.plu.edu/~ryandp/RAX.html (accessed June 29, 2014).

3. "Reed Boats," Wikipedia. http://en.wikipedia.org/wiki/Reed_boat (accessed June 28, 2014).

4. Richard Winsor, "Marine Technology," http://www.billbrouard.com/boat.htm (accessed June 28, 2014).

5. Andrew Lawler, "Report of Oldest Boat Hints at Early Trade Routes," *Science*, 296, no. 5574 (June 7, 2002), 1791-1792 DOI: 10.1126/science.296.5574.1791.

6. "Reed Boats," Wikipedia.

7. Dollinger, "Papyrus."

8. Winsor, "Marine Technology."

9. André Dollinger, "Ancient Egyptian ships and boats: The archaeological evidence," *An introduction to the history and culture of Pharaonic Egypt*, http://www.reshafim.org.il/ad/egypt/timelines/topics/navigation.htm (accessed June 28, 2014).

10. J.M. Allen, "History of Reed Ships," http://www.atlantisbolivia.org/areedboathistory.htm (accessed June 28, 2014).

11. "Thor Heyerdahl Expeditions and Archaeology of the Pacific Peoples," http://www.greatdreams.com/thor.htm (accessed June 28, 2014).

Chapter 5

1. James Hornell, *Water Transport: Origins and Early Evolution*, Revised ed.; introduction by Basil Greenhill. (Newton Abbot, U.K.: David & Charles, 1970), 187.

2. Lincoln P. Paine, *The Sea & Civilization: A Maritime History of the World* (New York: A. Knopf, 2013), 42.

3. Paine, *The Sea & Civilization*, 42.

4. Paine, *The Sea & Civilization*, 44–45.

5. Paine, *The Sea & Civilization*, 94.

6. Mark Cartwright, "Trireme," Ancient History Encyclopedia (May 31, 2012): http://www.ancient.eu/trireme/ (accessed September 9, 2015).

7. "Galley," *Encyclopaedia Britannica*, http://www.britannica.com/EBchecked/topic/224325/galley (accessed June 28, 2014).

8. John H. Leinhard, "The Last Galleys," *Engines of Our Ingenuity*, http://www.uh.edu/engines/epi303.htm (accessed June 28, 2014).

9. Paine, *The Sea &Civilization*, 112.

10. "History of Arms and Armour: Greek Fire," *History World*, http://www.historyworld.net/wrldhis/PlainTextHistories.asp?gtrack=pthc&ParagraphID=dlw#dlwGreek fire: 674 (accessed June 28, 2014).

Chapter 6

1. Paine, *The Sea & Civilization*, 42–43.
2. Paine, *The Sea & Civilization*, 255.
3. John Frayler, "The Great Age of Duck," *Pickled Fish and Salted Provisions: Historical Musings from Salem Maritime NHS*, 7, no. 4 (September 2005), http://www.nps.gov/sama/historyculture/upload/Vol7no4duck.pdf (accessed October 7, 2014).
4. "What are sails made out of?" Answers.com, http://wiki.answers.com/Q/What_are_sails_made_out_of (accessed July 23, 2014).
5. "Clipper Ship," *Encyclopaedia Britannica*, http://www.britannica.com/EBchecked/topic/121871/clipper-ship (accessed July 23, 2014).
6. Paine, *The Sea & Civilization*, 521
7. "Clipper Ships: Greyhounds of the Sea," http://www.anmm.gov.au/site/page.cfm?u=1307 (accessed July 24, 2014).
8. "Rolls-Royce Revives Age of Sail to Beat Fuel-Cost Surge," Bloomberg.com, http://www.bloomberg.com/news/2013-07-10/rolls-royce-revives-age-of-sail-to-beat-fuel-cost-surge-freight.html (accessed July 23, 2014).
9. "Sailing at the Touch of a Button," *Low-Tech Magazine*, http://www.lowtechmagazine.com/2009/04/sailing-ships-large-crew-automated-control.html (accessed July 23, 2014).
10. Stuart Fox, "How Do Solar Sails Work?" *LiveScience* (May 17, 2010): http://www.livescience.com/32593-how-do-solar-sails-work-.html (accessed April 14, 2015).
11. "Japanese Spacecraft Deploys Solar Sail," Space.com (June 11, 2010): http://www.space.com/8584-japanese-spacecraft-deploys-solar-sail.html (accessed April 14, 2015).
12. "Windsurfing on a Wicked World," NASA.gov, http://www.nasa.gov/directorates/spacetech/home/feature_windsurfing.html (accessed July 23, 2014).

Chapter 7

1. "History of Snow Sports," Fédération International de Ski, http://www.fis-ski.com/inside-fis/about/fis-history/snowsports/ (accessed October 19, 2014).
2. Suemedha Sood, "Where Did Skiing Come From?" BBC.com (December 22, 2010): http://www.bbc.com/travel/blog/20101221-travelwise-where-did-skiing-come-from (accessed October 20, 2014).
3. "A History of Skis," *National Geographic*, http://ngm.nationalgeographic.com/2013/12/first-skiers/ski-history-interactive (accessed October 20, 2014).
4. Sood, "Where Did Skiing Come From?"
5. Mike Doyle, "The History of Downhill and Cross Country Skiing," About.com, http://skiing.about.com/od/downhillskiing/a/skiinghistory.htm (accessed October 20, 2014).
6. "A History of Skis."
7. "Skiing History," Holmenkollen Ski Museum, https://web.archive.org/web/20150225150441/http://www.holmenkollen.com/eng/The-Ski-Museum/Skiing-history.
8. Morten Lund and Seth Masia, "A Short History of Skis," International Skiing History Association, https://skiinghistory.org/history/short-history-skis-0 (accessed October 20, 2014).
9. Sood, "Where Did Skiing Come From?"

Chapter 8

1. Marsha A. Levine, "Domestication, Breed Diversification and Early History of the Horse," Havemeyer Foundation, http://research.vet.upenn.edu/HavemeyerEquineBehaviorLab-HomePage/ReferenceLibraryHavemeyerEquineBehaviorLab/HavemeyerWorkshops/HorseBehaviorandWelfare1316June2002/HorseBehaviorandWelfare2/Domestication-BreedDiversificationandEarlyHis/tabid/3127/Default.aspx (accessed April 2, 2016).

2. Laura Klappenbach, "Domestication of Horses: The Relationship Between Horses and Humans," About.com, http://animals.about.com/od/hoofedmammals/a/domesticationof.htm (accessed December 10, 2014).

3. Alan Outram, "Horse domestication in the Botai Culture, Eneolithic Kazakhstan," University of Exeter, https://humanities.exeter.ac.uk/archaeology/research/projects/title_84579_en.html (accessed December 10, 2014).

4. "Ancient Civilizations - Mesopotamia: Animals," British Museum, http://www.mesopotamia.co.uk/staff/resources/background/bg27/home.html (accessed December 10 2014).

5. "Western Zhou Chariot Burial Pit," TravelChinaGuide.com, http://www.travelchinaguide.com/attraction/shaanxi/xian/westernzhou.htm (accessed December 10, 2014).

6. Jennie Cohen, "Horse Domestication Happened Across Eurasia, Study Shows," *History.com* (January 30, 2012), http://www.history.com/news/horse-domestication-happened-across-eurasia-study-shows (accessed December 10, 2014)

7. Levine, "Domestication."

8. "Draft Animal," *Encyclopaedia Britannica*, http://www.britannica.com/EBchecked/topic/170716/draft-animal (accessed December 10, 2014).

9. "War Elephants," HellenicaWorld.com, http://www.hellenicaworld.com/Greece/LX/en/WarElephant.html (accessed April 2, 2016).

10. "Deadliest Warrior – War Elephant," Wikia, http://deadliestwarrior.wikia.com/wiki/Elephant (accessed April 29, 2015).

11. Kevin Knodell, "Battle of the Dumbos: Elephant Warfare from Ancient Greece to the Vietnam War," War Is Boring, https://medium.com/war-is-boring/battle-of-the-dumbos-elephant-warfare-from-ancient-greece-to-the-vietnam-war-ca62af225917 (accessed April 29, 2015).

Chapter 9

1. Jŭs Turk, "Mysterious pile-dwellers, a revelation about prehistoric people in the Ljubljansko barje," Slovenia.si (April 2011): http://www.slovenia.si/slovenia/history/earliest-traces/mysterious-pile-dwellers-a-revelation-about-prehistoric-people-in-the-ljubljansko-barje/ (accessed December 10, 2014).

2. "Frequently Asked Questions," Swann Mountain Llama Trekking, http://www.llama-treksmontana.com/frequently-asked-questions (accessed December 10, 2014).

3. Tudor Raiciu, "History of the Wheel," Autoevolution.com (June 2, 2009): http://www.autoevolution.com/news/history-of-the-wheel-7334.html (accessed September 16, 2015).

4. "Bearing (mechanical)," Wikipedia, https://en.wikipedia.org/wiki/Bearing_%28mechanical%29 (accessed September 16, 2015).

Chapter 10

1. John Noble Wilford, "Remaking the Wheel: Evolution of the Chariot," *The New York Times Archive* (February 22, 1994): http://www.nytimes.com/1994/02/22/science/remaking-the-wheel-evolution-of-the-chariot.html (accessed October 10, 2014).

2. Jimmy Dunn (writing as Troy Fox), "The Chariot in Egyptian Warfare," TourEgypt.Net, http://www.touregypt.net/featurestories/chariots.htm (accessed October 10, 2014).

3. "Making History."

4. "Making History."

5. André Dollinger, "The Chariot," An introduction to the history and culture of Pharaonic Egypt, http://www.reshafim.org.il/ad/egypt/timelines/topics/chariot.htm (accessed October 8, 2014).

6. Dunn, "The Chariot in Egyptian Warfare."

7. Dunn, "The Chariot in Egyptian Warfare."

8. "Making History."

9. "Building the Pharaoh's Chariot," NOVA video (run time 52:52, originally aired February 16, 2013): http://www.pbs.org/wgbh/nova/ancient/pharaoh-chariot.html (accessed October 8, 2014).

Chapter 11

1. Kris De Decker, "Aerial ropeways: automatic cargo transport for a bargain" (January 2011): *Low-Tech Magazine*, http://www.lowtechmagazine.com/2011/01/aerial-ropeways-automatic-cargo-transport.html (accessed December 7, 2014).
2. "Aerial ropeways."
3. Sood, "Where Did Skiing Come From?"
4. Brian Yarvin, "A brief history of ski lifts and cable cars," The Bend Bulletin (last updated November 19, 2013): http://www.bendbulletin.com/news/1351807-151/a-brief-history-of-ski-lifts-and-cable (accessed December 7, 2014).
5. Daniel Engber, "Who Made That Ski Lift?" *New York Times Magazine* (February 21, 2014): http://www.nytimes.com/2014/02/23/magazine/who-made-that-ski-lift.html?_r=0 (accessed December 7, 2014).
6. "About Ropeways," Internet archive, http://web.archive.org/web/20060904183704/http://www.mines.edu/library/ropeway/about_ropeways.html (accessed June 6, 2015).
7. "Amazing Aerial Tramways of the World," Kuriositas, (August 12,2014): http://www.kuriositas.com/2011/02/amazing-aerial-tramways-of-world.html (accessed December 10, 2014).

Chapter 12

1. "Principal Deserts of the World," FactMonster.com, http://www.factmonster.com/ipka/a0778851.html (accessed May 26, 2015).
2. Melissa Gish, *Living Wild: Camels* (Mankato, MN, Creative Education, 2013), 10.
3. Gish, *Living Wild: Camels*, 11.
4. Gish, *Living Wild: Camels*, 19.
5. Gish, *Living Wild: Camels*, 15.
6. Gish, *Living Wild: Camels*, 16.
7. Gish, *Living Wild: Camels*, 16.
8. Gish, *Living Wild: Camels*, 10.
9. "Caravan: Desert Transport," *Encyclopaedia Britannica*, http://www.britannica.com/EBchecked/topic/94606/caravan (accessed May 26, 2015).
10. "Caravan: Desert Transport."
11. Chris Rainier, "In Sahara, Salt-Hauling Camel Trains Struggle On," *National Geographic News* (May 28, 2003): http://news.nationalgeographic.com/news/2003/05/0528_030528_saltcaravan.html (accessed May 22, 2015).
12. "Camel Trains in the Desert," ChinaVista.com, http://www.chinavista.com/experience/camel/camel.html (accessed May 12, 2015).
13. John Pilkington, "Dying trade of the Sahara camel train," BBC News (last updated October 21, 2006): http://news.bbc.co.uk/2/hi/programmes/from_our_own_correspondent/6070400.stm (accessed May 22, 2015).
14. William J. Bernstein, *A Splendid Exchange: How Trade Shaped the World* (New York: Atlantic Monthly Press, 2008), 75.
15. Colin Thurbron, *Shadow of the Silk Road* (HarperCollins Publishers, New York, 2007).
16. Gish, *Living Wild: Camels*, 32.
17. "Afghan Cameleers in Australia," Australia.gov.au, http://www.australia.gov.au/about-australia/australian-story/afghan-cameleers (accessed May 27, 2015).

Chapter 13

1. Thom Swan, "'Marche': Sledge Dogs in the North West Fur Trade," Stardancer Historical Freight Dogs, http://www.tworiversak.com/sleddoghx1.htm (accessed November 20, 2014).

2. Swan, "Marche."

3. Swan, "Marche."

4. Swan, "Marche."

5. Don Bowers, "History, the World was Changing…," Iditarod.com (updated 2012): http://iditarod.com/about/history/ (accessed November 23, 2014).

6. Bowers, "History."

7. Christopher Klein, "The Sled Dog Relay that Inspired the Iditarod," History.com (March 10, 2014): http://www.history.com/news/the-sled-dog-relay-that-inspired-the-iditarod (accessed November 20, 2014).

Chapter 14

1. "Marco Polo, Explorer, Journalist (c. 1254–1324)," Biography.com, http://www.biography.com/people/marco-polo-9443861 (accessed November 21, 2014).

2. Paine, *The Sea & Civilization*, 173.

3. Paine, *The Sea & Civilization*, 177.

4. Paine, *The Sea & Civilization*, 358.

5. Paine, *The Sea & Civilization*, 193.

6. Kallie Szczepanski, "Zheng He's Treasure Ships," About.com, http://asianhistory.about.com/od/china/p/Zheng-Hes-Treasure-Ships.htm (accessed November 21, 2014).

7. Evan Hadingham, "Ancient Chinese Explorers," NOVA (posted January 16, 2001): http://www.pbs.org/wgbh/nova/ancient/ancient-chinese-explorers.html (accessed November 21, 2014).

8. Hadingham, "Ancient Chinese Explorers."

9. Gavin Menzies, *1421: The Year China Discovered America* (New York: HarperCollins Publishers, 2002), 270–272.

10. Menzies, *1421*, 37.

11. P.J. Rivers, "Will the real Gavin Menzies please stand up!" The "1421" Myth Exposed http://www.1421exposed.com/ (accessed December 3, 2014).

12. Richard Gunde, "Zheng He's Voyages of Discovery," UCLA Asia Institute, http://web.archive.org/web/20140320101436/http://www.international.ucla.edu/asia/news/article.asp?parentid=10387 (accessed December 29, 2015).

13. Michael L. Bosworth, "The Rise and Fall of 15th Century Chinese Sea Power," http://web.archive.org/web/20070529193206/http://www.cronab.demon.co.uk/china.htm (accessed December 19, 2015).

14. Hadingham, "Ancient Chinese Explorers."

15. Paine, *The Sea & Civilization*, 390.

Chapter 15

1. "History of Wheelchairs," Wheelchair Information, http://www.wheelchair-information.com/history-of-wheelchairs.html (accessed November 28, 2014).

2. "History of the Wheelchair," Mobility Scooters Otago, http://www.mobilityscooters.co.nz/history/wheelchairs (accessed November 28, 2014).

3. Brian Woods, "History of the Wheelchair," *Encyclopaedia Britannica*, http://www.britannica.com/EBchecked/topic/1971423/history-of-the-wheelchair (accessed November 28, 2014).

4. Mary Bellis, "History of the Wheelchair," About.com, http://inventors.about.com/od/wstartinventions/a/wheelchair.htm (accessed November 28, 2014).

5. Bellis, "History of the Wheelchair."

6. Woods, "History of the Wheelchair."

7. "History of Wheelchairs," Wheelchair Information (website).

Chapter 16

1. Bamber Gascoigne, "History of Transport and Travel, Carriages: 17th Century," History World (from 2001, ongoing): http://www.historyworld.net/wrldhis/PlainTextHistories. asp?groupid=1972&HistoryID=ab79&gtrack=pthc (accessed December 7, 2014).

2. Gascoigne, "History of Transport and Travel."

3. Frank E. Huggett, *Carriages at Eight: Horse-drawn Society in Victorian and Edwardian Times* (New York: Charles Scribner's Sons, 1979), 27–32.

4. Huggett, *Carriages at Eight*, 55.

5. Huggett, *Carriages at Eight*, 32.

6. Huggett, *Carriages at Eight*, 35.

7. Huggett, *Carriages at Eight*, 26.

8. Huggett, *Carriages at Eight*, 13.

9. Huggett, *Carriages at Eight*, 85.

10. Huggett, *Carriages at Eight*, 131–132.

11. Huggett, *Carriages at Eight*, 115–117.

12. Mara Gay, "Carriage Horses, Now Controversial, Have a Long City History," *Wall Street Journal, Metropolis* (March 24, 2014): http://blogs.wsj.com/metropolis/2014/03/24/ carriage-horses-now-controversial-have-a-long-city-history/ (accessed December 8, 2014).

13. Huggett, *Carriages at Eight*, 17.

14. Zainab Abdulaziz, "History on wheels: 5 things to know about the queen's new carriage," Today.com (June 4, 2014): http://www.today.com/news/history-wheels-5-things-know-about-queens-new-carriage-2D79756846 (accessed December 7, 2014).

15. "Grand Tour," Wikipedia, http://en.wikipedia.org/wiki/Grand_Tour (accessed April 30, 2015).

Chapter 17

1. Kelly L. Wheeler, "The History of Yachting," ezinearticles.com (submitted on April 2, 2008): http://ezinearticles.com/?The-History-of-Yachting&id=1084031 (accessed December 7, 2014).

2. "'Yacht' A Little History," Stability Yachts, http://www.stabilityyachts.com/yacht.html (accessed December 7, 2014).

3. L.V. Anderson, "What Is a Yacht? And when did owning one become a symbol of wealth?" Slate.com (December 13, 2012): http://www.slate.com/articles/life/luxury_ explainer/2012/12/what_is_a_yacht_and_why_are_they_associated_with_luxury.html (accessed December 7, 2014).

4. Anderson, "What Is a Yacht?"

5. "The History of Luxury Yachts," Luxury Yachts Charter, http://www.megayachtscharter. com/en/general-information/history-of-luxury-yachts/ (accessed December 8, 2014).

6. "The History of Luxury Yachts."

7. "History of the America's Cup Races," 12 Meter Charter, http://12metercharters.com/ americas-cup-race-history (accessed December 8, 2014).

Chapter 18

1. "Taxicab," Wikipedia, http://en.wikipedia.org/wiki/Taxicab.

2. Mary Bellis, "Hailing – History of the Taxi," About.com, http://inventors.about.com/od/ tstartinventions/a/taxi.htm (accessed December 8, 2014).

3. "June 1654, An Ordinance for the Regulation of Hackney-Coachmen in London and the places adjacent," British History Online, http://www.british-history.ac.uk/report. aspx?compid=56562 (accessed December 7, 2014).

4. Cindy Crank, Transport and Carriages in the Victorian Era (1837–1901), Horses and History Throughout the Ages (blog, posted May 2, 2011): https://horsesandhistory.wordpress.com/about/ (accessed December 7, 2014).

5. Crank, "Transport and Carriages."

6. Huggett, *Carriages at Eight*, 100.

7. Bellis, "Hailing – History of the Taxi."

8. Bellis, "Hailing – History of the Taxi."

9. Bellis, "Hailing – History of the Taxi."

10. "Taxi History," PBS.org, https://web.archive.org/web/20150310001737/http://www.pbs.org/wnet/taxidreams/history/ (accessed April 2, 2016).

11. "Taxi History," PBS.org.

12. "Taxi Drivers and Chauffeurs," *U.S. Dept. of Labor, Bureau of Labor Statistics, Occupational Outlook Handbook*, http://www.bls.gov/ooh/transportation-and-material-moving/taxi-drivers-and-chauffeurs.htm (accessed December 7, 2014).

Chapter 19

1. Greg Goebel, "The Invention of the Submarine," Internet archive, https://web.archive.org/web/20020820171807/http://www.vectorsite.net/twsub1.html (accessed June 8, 2015).

2. Brett McLaughlin, "Cornelius Drebbel, Inventor of the Submarine," Dutchsubmarines.com, http://www.dutchsubmarines.com/specials/special_drebbel.htm (accessed April 27, 2015).

3. Goebel, "The Invention of the Submarine."

4. Paul Marks, "From sea to sky: Submarines that fly," NewScientist.com (July 5, 2010): http://www.newscientist.com/article/mg20727671.000-from-sea-to-sky-submarines-that-fly.html#.VId_8V50xjo (accessed March 28, 2014).

5. "DARPA Plans to Develop 'Flying Submarine,'" NavalTechnology.com (July 8, 2010): http://www.naval-technology.com/news/news89904.html (accessed March 28, 2014).

6. DeepFlight corporate website, http://www.deepflight.com/ (accessed May 12, 2015).

7. Triton Subs corporate website, http://tritonsubs.com/# (accessed May 12, 2015).

Chapter 20

1. "Coaching History," Regency Collection, http://homepages.ihug.co.nz/~awoodley/carriage/history.html (accessed December 12, 2014).

2. "Coaching History."

3, "Stagecoach Travel," *Dictionary of American History* (2003): accessed at Encyclopedia.com, http://www.encyclopedia.com/doc/1G2-3401804002.html (accessed December 12, 2014).

4. Elizabeth Larson, "The Concord Coach," Over-Land.com, http://www.over-land.com/ccoach.html (accessed November 29, 2014).

5. "Stagecoach Travel."

6. "Stagecoach Travel."

Chapter 21

1. Randy Alfred, "1662: The Bus Starts Here…in Paris," Wired.com (March 18, 2008): http://archive.wired.com/science/discoveries/news/2008/03/dayintech_0318 (accessed December 12, 2014).

2. "A Short History of Public Transport in Greater Manchester," Greater Manchester Museum of Transport (revised May 2004): https://web.archive.org/web/20150707163229/http://www.gmts.co.uk/explore/history/history.html (accessed April 2, 2016).

3. "Stanislas Baudry et les omnibus de Nantes en 1825," French Influence (blog, February 25, 2011): http://frenchinfluence.over-blog.fr/article-stanislas-baudry-et-les-omnibus-de-nantes-en-1825-68083889.html (accessed December 20, 2014).

4. Huggett, *Carriages at Eight*, 104–105.
5. Huggett, *Carriages at Eight*, 105–109.
6. "A Brief History of the Bus," Bus Stuff, http://www.bus-stuff.co.uk/2011/01/brief-history-of-bus.html (accessed December 12, 2014).
7. Jordan Golson, "A Giant Charger That Juices Up Electric Buses in Three Minutes," Wired.com (October 10, 2014): http://www.wired.com/2014/10/giant-charger-juices-electric-buses-three-minutes/ (accessed December 12, 2014).
8. "Electric Buses: Rapid Charging in Vienna," Siemens.com, http://www.siemens.com/innovation/en/home/pictures-of-the-future/mobility-and-motors/electric-mobility-electric-buses.html (accessed April 2, 2016).
9. "A Brief History of the Bus."
10. "A Brief History of the Bus."

Chapter 22

1. Rick Falkvinge, "The Red Flag Act of 1865," TorrentFreak (June 26, 2011): https://torrent-freak.com/the-red-flag-act-of-1865-110626/ (accessed September 17, 2014).
2. Mary Bellis, "History of the Automobile," About.com, http://inventors.about.com/library/weekly/aacarssteama.htm (accessed September 15, 2014).
3. William W. Bottorff, "What Was the First Car? A Quick History of the Automobile for Young People," https://web.archive.org/web/20140701050734/http://www.ausbcomp.com/~bbott/cars/carhist.htm (accessed June 8, 2015).
4. Bellis, "History of the Automobile."
5. Mary Bellis, "Gottlieb Daimler," About.com, http://inventors.about.com/od/dstartinventors/a/Gottlieb_Daimler.htm (accessed September 15, 2014).
6. Ron Dick and Dan Patterson, *Aviation Century: The Early Years*, (Erin, Ontario: Boston Mills Press, 2003), 17.
7. David Burgess Wise, *The Illustrated History of Automobiles* (New York: A & W Publishers, 1980), 14.
8. Tom D. Crouch, *Wings: A History of Aviation from Kites to the Space Age* (Washington, DC: Smithsonian National Air and Space Museum, 2003), 240.
9. "Automobiles," History.com, http://www.history.com/topics/automobiles (accessed September 17, 2014).
10. *Cycle and Automobile Journal*, quoted in "Automobiles," History.com http://www.history.com/topics/automobiles (accessed September 17, 2014).
11. Michael McGerr, *A Fierce Discontent: The Rise and Fall of the Progressive Movement in America* (New York: Oxford University Press, 2003), 230.
12. "First American Oil Well," American Oil & Gas Historical Society, http://aoghs.org/oil-amanac/american-oil-history/ (accessed May 12, 2015).
13. George Constable and Bob Somerville, *A Century of Innovation: Twenty Engineering Achievements That Transformed Our Lives* (Washington, DC: National Academy of Engineering, 2003): accessed at: http://www.greatachievements.org/?id=2965 (accessed May 12, 2015).
14. "How We Use Energy: Transportation," The National Academies, What You Need to Know About Energy, http://needtoknow.nas.edu/energy/energy-use/transportation/ (accessed May 12, 2015).

Chapter 23

1. "The History of Hot Air Ballooning," eballoon.org, http://www.eballoon.org/history/history-of-ballooning.html (accessed December 9, 2014).
2. "The History of Hot Air Ballooning," Arizona Balloon Club, http://www.arizonaballoon-club.org/history_of_ballooning.htm (accessed December 9, 2014).

3. Julian Nott, "The Extraordinary Nazca Prehistoric Balloon," http://juliannott.com/nazca/ (accessed December 9, 2014).

4. Liesl Clark, "A Short History of Ballooning," PBS.org (December 2, 1997): http://www.pbs.org/wgbh/nova/space/short-history-of-ballooning.html.

5. "The History of Hot Air Ballooning."

6. Clark, "A Short History of Ballooning."

7. Jason Golomb, "Nasca Lines – The Sacred Landscape," *National Geographic*, http://science.nationalgeographic.com/science/archaeology/nasca-lines/ (accessed December 9, 2014).

8. "Nasca Lines Theories," bibliotecapleyades.net, http://www.bibliotecapleyades.net/nazca/esp_lineas_nazca_2.htm (accessed December 9, 2014).

Chapter 24

1. "Industrial Revolution," History.com (2009): http://www.history.com/topics/industrial-revolution (accessed July 24, 2014).

2. "Steamships," The Open Door Web Site, http://www.saburchill.com/history/chapters/IR/033.html (accessed July 23, 2014).

3. "Steamships."

4. "A History of Steamboats," U.S. Army Corps of Engineers, http://www.sam.usace.army.mil/Portals/46/docs/recreation/OP-CO/montgomery/pdfs/10thand11th/ahistoryof-steamboats.pdf (accessed July 23, 2014).

5. "Canal History," New York State Canal Corporation, http://www.canals.ny.gov/history/history.html (accessed October 9, 2014).

6. Daniel Othfors and Henrik Ljungström, "The Great Ocean Liners" (updated June 24, 2014): http://www.thegreatoceanliners.com/index2.html (accessed August 8, 2014).

7. Othfors and Ljungström, "The Great Ocean Liners."

8. "Steamships," The Open Door Web Site.

9. "Ships and Steam Power," Royal Museums Greenwich, http://www.rmg.co.uk/explore/sea-and-ships/facts/ships-and-seafarers/steam-power (accessed June 2, 2014).

10. "A History of Steamboats," U.S. Army Corps of Engineers.

Chapter 25

1. Tim Frew, *Locomotives: From the Steam Locomotive to the Bullet Train* (New York: Mallard Press, 1990), 22.

2. "Steam Train Anniversary Begins," BBC News (updated February 21, 2004): http://news.bbc.co.uk/2/hi/uk_news/wales/3509961.stm (accessed June 3, 2014).

3. L.T.C. Rolt, "Richard Trevithick," *Encyclopaedia Britannica* (updated May 18,2014): http://www.britannica.com/EBchecked/topic/604469/Richard-Trevithick (accessed September 9, 2014).

4. Frew, *Locomotives*, 29.

5. Frew, *Locomotives*, 37.

6. Frew, *Locomotives*, 69–70.

7. Frew, *Locomotives*, 81–84.

8. Frew, *Locomotives*, 128.

9. Frew, Locomotives, 53–56.

10. William C. Kashatus, "Legend of Lackawanna Railroad's Phoebe Snow lives on," Citizensvoice.com (October 23, 2011): http://citizensvoice.com/arts-living/legend-of-lackawanna-railroad-s-phoebe-snow-lives-on-1.1220950 (accessed April 28, 2015).

Chapter 26

1. Arallyn Primm, "By Land and Sea: The History of Amphibious Vehicles," *Mental Floss* (May 28, 2013): http://mentalfloss.com/article/50725/land-and-sea-history-amphibious-vehicles (accessed October 22, 2014).

2. Charles F. Kettering, "Short Stories of Science and Invention, A Collection of Radio Talks," Today in Science and History website (reprint March 1959): http://todayinsci.com/stories2/story013.htm (accessed June 9, 2015).

3. Steven Lubar, "Was This America's First Steamboat, Locomotive, and Car?" *American Heritage's Innovation & Technology*, 21, no: 4 (Spring 2006): http://www.innovationgateway.org/content/was-america%E2%80%99s-first-steamboat-locomotive-and-car-0 (accessed October 22, 2014).

4. Primm, "By Land and Sea."

5. "Channel Crossing Biography," Dover Museum, http://www.dovermuseum.co.uk/Information-Resources/HPAC-Biographical/Channel-Crossing-Biography.aspx (accessed April 28, 2015).

6. "A Brief History of Amphibious Vehicles," Unique Cars and Parts, http://www.uniquecarsandparts.com.au/history_amphibians.htm (accessed October 22, 2014).

7. Primm, "By Land and Sea."

8. "DUKW (Duck)," U.S. Army Transportation Museum, http://web.archive.org/web/20080725043610/http:/www.transchool.eustis.army.mil/Museum/DUKW.htm (accessed October 22, 2014).

9. "A history of amphibious cars, from the Amphicar to the Gibbs Sport Quadski," *New York Daily News* (October 22, 2012): http://www.nydailynews.com/autos/unusual-amphibious-cars-article-1.1189217 (accessed October 19, 2014).

10. Phil Berg, "Swimming with Cars: 9 Amphibious Vehicles," *Popular Mechanics*, http://www.popularmechanics.com/cars/news/pictures/swimming-with-cars-9-amphibious-vehicles#slide-1 (accessed October 19, 2014).

11. Primm, "By Land and Sea."

Chapter 27

1. "American Western Migration: Wagon Trains and Covered Wagons," Historybits, http://www.historybits.com/west-wagon-trains.htm (accessed July 12, 2014).

2. "American Western Migration."

3. George R. Stewart, "The Prairie Schooner Got Them There," *American Heritage Magazine*, 13, no: 2 (February 1962): http://www.americanheritage.com/content/prairie-schooner-got-them-there (accessed July 12, 2014).

4. Stewart, "The Prairie Schooner Got Them There."

5. Stephenie Flora, ed., "The Covered Wagon," oregonpioneers.com (2007): http://www.oregonpioneers.com/wagon.htm (accessed July 10, 2014).

6. Stewart, "The Prairie Schooner Got Them There."

7. Stewart, "The Prairie Schooner Got Them There."

8. "Prairie Schooner," *Encyclopedia Britannica*, http://www.britannica.com/EBchecked/topic/473853/prairie-schooner.

9. Stewart, "The Prairie Schooner Got Them There."

Chapter 28

1. Maria C Collins, "The Golden Age of Ocean Liners, 1900–1914," Humanities 360 (March 11, 2013): https://web.archive.org/web/20140630172225/http://www.humanities360.com/index.php/the-golden-age-of-ocean-liners-1900-1914-1975/ (accessed June 9, 2015).

2. Othfors, "The Great Ocean Liners."

3. "Liners to America," Smithsonian National Museum of American History, http://amhistory.si.edu/onthewater/exhibition/5_2.html (accessed July 12, 2014).

4. "Liners to America."

5. "Ocean Liners," Wikipedia, http://en.wikipedia.org/wiki/Ocean_liner (accessed July 12, 2014).

6. Collins, "The Golden Age of Ocean Liners."

7. Gregory J. Norris, "Evolution of Cruising," *Cruise Travel*, 3, no: 3 (November–December 1981), 28, accessed at https://books.google.com/books?id=IDEDAAAAMBAJ&pg=PA 28&lpg=PA28&dq=norris+evolution+of+cruising&source=bl&ots=eSoZ4f4jD9&sig= AJLnQj2eyZKhluoqf1uJehamahU&hl=en&sa=X&ei=V0t3VbbeHcyKsAXalICoCw&v ed=0CCUQ6AEwAQ#v=onepage&q=norris%20evolution%20of%20cruising&f=false (accessed July 12, 2014).

8. "Ocean Liners," Wikipedia.

Chapter 29

1. "Jules Henri Giffard's Steam Airship," The History Forum, http://www.thehistoryforum. com/airships/henri_giffard/ (accessed August 2, 2014).

2. R.G. Grant, *Flight, The Complete History: 100 Years of Aviation* (New York: DK Publishing, 2002, 2007), 57.

3. Crouch, *Wings: A History of Aviation*, 605.

4. Grant, *Flight, The Complete History*, 57.

5. Grant, *Flight, The Complete History*, 157.

6. Crouch, *Wings: A History of Aviation*, 294.

7. "Hindenburg Flight Schedule," Airships.net, http://www.airships.net/hindenburg/flight-schedule (accessed August 2, 2014).

8. Crouch, *Wings: A History of Aviation*, 294–295.

Chapter 30

1. "Elevators & Escalators – History of the Elevator," Mitsubishielectric.com, http://www. mitsubishielectric.com/elevator/overview/elevators/history.html (accessed May 12, 2014).

2. "About Elevators," Otisworldwide.com, http://www.otisworldwide.com/pdf/AboutEleva-tors.pdf (accessed May 12, 2014).

3. "About Elevators."

4. Mary Bellis, "History of the Elevator," About.com, http://inventors.about.com/od/ estartinventions/a/Elevator.htm (accessed May 11, 2014).

5. "Elevators & Escalators – History of the Elevator."

6. Ralph M. Newman, "Elevator Door Operators: Then and Now: Attendants of yesteryear have given way to high-tech, automated solutions," Columbuselevator.com, http://www. columbiaelevator.com/clientuploads/pdf/CE-Elevator-Door-Operators.pdf (accessed May 12, 2014).

7. "About Elevators."

8. Leon Neyfakh, "How the elevator transformed America," *The Boston Globe* (March 2, 2014): http://www.bostonglobe.com/ideas/2014/03/02/how-elevator-transformed-amer-ica/b8u17Vx897wUQ8zWMTSvYO/story.html (accessed May 12, 2014).

9. "Otis Passenger Elevator – Impact," Otis Elevator (blog): http://otiselevator.umwblogs.org/ impact/ (accessed April 29, 2015).

10. Neyfakh, "How the elevator transformed America."

Chapter 31

1. Benson Bobrick, *Labyrinths of Iron: A History of the World's Subways* (New York: Newsweek Books, 1981), 144.

2. Tracy V. Wilson, "How Subways Work," How Stuff Works, http://science.howstuffworks. com/engineering/civil/subway.htm (accessed September 6, 2014).

3. Bobrick, *Labyrinths of Iron*, 135.

4. Bobrick, *Labyrinths of Iron*, 96.

5. Bobrick, *Labyrinths of Iron*, 103.

6. Ed Kirby, "Inventor Frank Sprague," *Seldom Told Tales of Sharon, Book Three* (Sharon, CT: Sharon Historical Society), accessed at https://web.archive.org/web/20131219020028/http://www.sharonhist.org/inventor-frank-sprague.pdf (accessed June 10, 2015).
7. Bobrick, *Labyrinths of Iron*, 314.
8. Bobrick, *Labyrinths of Iron*, 280.
9. Bobrick, *Labyrinths of Iron*, 285.
10. "New York City Subway in popular culture explained," http://everything.explained.today/New_York_City_Subway_in_popular_culture/ (accessed December 28, 2015).

Chapter 32
1. "A Quick History of Bicycles," Pedaling History Bicycle Museum, https://web.archive.org/web/20030716020048/http://www.pedalinghistory.com/PHhistory.html (accessed June 10, 2015).
2. Canada Science and Technology Museum, Bicycle Collection, http://www.sciencetech.tech-nomuses.ca/english/collection/cycles1.cfm accessed 10/7/14.
3. "A Quick History of Bicycles."
4. "A Quick History of Bicycles."
5. "A Quick History of Bicycles."
6. Pauline Weston Thomas, "Rational Dress Reform: Fashion History – Mrs. Bloomer," Fashion-Era.com, http://www.fashion-era.com/rational_dress.htm (accessed April 28, 2015).
7. David Hendricks, "The Possibility of Mobility: The Rise and Fall of the Bicycle in 19th Century America" (last updated January 7, 2010): http://xroads.virginia.edu/~ug02/hendrick/bikehome.html (accessed February 10, 2014).
8. "Bikeshare Cities," Bikeshare website, http://bikeshare.com/map/ (accessed February 10, 2014).
9. David Fiedler, "Bikes – An Illustrated History," About.com, http://bicycling.about.com/od/thebikelife/ss/History.htm (accessed February 10, 2014).

Chapter 33
1. Mary Bellis, "History of the Motorcycle," About.com, http://inventors.about.com/od/mstartinventions/a/motorcycle.htm (accessed August 22, 2014).
2. Tony Long, "This Day in Tech, 19th century, Aug. 30, 1885: Daimler Gives World First 'True' Motorcycle," *Wired* (August 30, 2011): http://www.wired.com/2011/08/0830daimler-first-true-motorcycle/ (accessed August 22, 2014).
3. William Harris, "How Motorcycles Work," How Stuff Works, http://auto.howstuffworks.com/motorcycle6.htm.
4. Harris, "How Motorcycles Work."
5. Richard Hough and L.J.K. Setright, *History of the World's Motorcycles* (New York: Harper & Row Publishers, 1966), 15–16.
6. Hough and Setright, *History of the World's Motorcycles*, 29.
7. Hough and Setright, *History of the World's Motorcycles*, 26.
8. Hough and Setright, *History of the World's Motorcycles*, 36.
9. Hough and Setright, *History of the World's Motorcycles*, 299.
10. Lisa Smedman, *From Boneshakers to Choppers: The Rip-Roaring History of Motorcycles* (Toronto, Ontario, Canada: Annick Press, Ltd., 2007), 2–3.
11. Smedman, *From Boneshakers to Choppers*, 72.
12. Hough and Setright, *History of the World's Motorcycles*, 153.

Chapter 34
1. Saikat Ray, "Fresh hope for hand-pulled rickshaws," *The Times of India* (December 23, 2011): http://timesofindia.indiatimes.com/city/kolkata/Fresh-hope-for-hand-pulled-rickshaws/articleshow/11212875.cms (accessed April 14, 2015).

2. Rory Boland, "Rickshaw and Its History." About.com, Travel, http://gohongkong.about.com/od/historyandcultureofhk/a/Rickshaw-A-Brief-History.htm (accessed April 14, 2015).

3. Emily Geens, "The History of the Rickshaw – Exploitation or Tradition?" *New Histories*, 2, no: 2.6 (April 14, 2011) http://newhistories.group.shef.ac.uk/wordpress/wordpress/the-history-of-the-rickshaw-exploitation-or-tradition/ (accessed April 14, 2015).

4. Margaret Simpson, "Japanese Rickshaw," Powerhouse Museum: Inside the Collection (blog, posted November 19, 2012): http://www.powerhousemuseum.com/insidethecollection/tag/jinrikisha/ (accessed April 14, 2015).

5. Simpson, "Japanese Rickshaw."

6. Simpson, "Japanese Rickshaw."

7. Boland, "Rickshaw – A brief history."

8. Calvin Trillin, "The Last Days of the Rickshaw," *National Geographic* (April 2008): http://ngm.nationalgeographic.com/2008/04/kolkata-rickshaws/calvin-trillin-text (accessed April 14, 2015).

9. Ray, "Fresh hope for hand-pulled rickshaws."

10. Trillin, "The Last Days of the Rickshaw."

11. Melissa Shales, "Durban's Ricksha boys – Durban's spectacular rickshaw runners. Few now remain but the Durban rickshaws are still a dramatic city icon." About.com, Travel, http://goafrica.about.com/od/durbankwazulunatal/a/durban-rickshaw.htm (accessed April 14, 2015).

Chapter 35

1. "Cruise Ship," Wikipedia, http://en.wikipedia.org/wiki/Cruise_ship (accessed May 27, 2014).

2. Douglas D. Slaten and Kiren Mitruka, "Cruise Ship Travel," *CDC Traveler's Health – Chapter 6, Conveyance and Transportation Issues*, http://wwwnc.cdc.gov/travel/yellowbook/2010/chapter-6/cruise-ship-travel.aspx (accessed May 27, 2014).

3. Slaten and Mitruka, "Cruise Ship Travel."

4. Viking River Cruises, www.vikingcruises.com, (accessed May 29, 2014).

5. John Hooper, "Italian Cruise Ship Fends Off Pirates with Gunfire," *The Guardian* (April 26, 2009): http://www.theguardian.com/world/2009/apr/26/italian-cruise-ship-pirates (accessed May 29, 2014).

Chapter 36

1. "What are the advantages and disadvantages of airships versus airplanes?" Quora.com, http://www.quora.com/What-are-the-advantages-and-disadvantages-of-airships-versus-airplanes (accessed April 28, 2015).

2. Ron Dick and Dan Patterson, *Aviation Century: The Early Years* (Erin, Ontario, Canada: Boston Mills Press, 2003), 18–20.

3. Curtis Pendergast and the editors of Time-Life Books, *The Epic of Flight: The First Aviators* (Alexandria, VA: Time-Life Books, 1981, 3rd printing revised 1986), 125.

4. Pendergast, *The Epic of Flight*, 67.

5. Crouch, *Wings: A History of Aviation*, 110.

6. Chad Upton, "Why Do Airplanes Fly at High Altitudes?" *Broken Secrets* (blog, posted February 23, 2010): http://brokensecrets.com/2010/02/23/why-do-airplanes-fly-at-high-altitudes/ (accessed April 18, 2015).

7. Crouch, *Wings: A History of Aviation*, 595–596.

8. Peter Gray, "How WWI changed aviation forever," *BBC News Magazine* (October 20, 2014): http://www.bbc.com/news/magazine-29612707 (accessed December 18, 2015).

9. "Military Aviation: Key Innovations," PBS.org, http://www.pbs.org/wnet/warplane/topicfeature.html (accessed December 28, 2015).

10. "Military Aviation: Key Innovations."

Chapter 37

1. Jane Allyson, "The History of Snowmobiles, *Sporting Life 360* (January 22, 2008): https://web.archive.org/web/20140706025723/http://www.sportinglife360.com/index.php/the-history-of-snowmobiles-56187/ (accessed December 30, 2015).
2. Stephen Campbell, "History of Snowmobiling," Northern Timber Cruisers Snowmobile Club, http://www.millinocket-maine.net/history-of-snowmobiling.htm (accessed November 13, 2014).
3. Campbell, "History of Snowmobiling."
4. Dixie Allan, "Snowmobiles," About.com, http://webclipart.about.com/od/seasonsclipart/ss/Snowmobiles.htm (accessed November 13, 2014).
5. Campbell, "History of Snowmobiling."
6. "History of Snowmobiles," *Wisconsin Snowmobile Education Course*, https://www.snowmobile-ed.com/wisconsin/studyGuide/History-of-Snowmobiles/50105101_700066571 (accessed November 13, 2014).
7. BRP.com, http://www.brp.com/en-us (accessed November 13, 2014).
8. "Snowmobiling Fact Book," *International Snowmobile Manufacturers' Association*, https://web.archive.org/web/20150523210556/http://snowmobile.org/snowmobilefacts.asp (accessed April 2, 2016).
9. Roger Lohr, "When snowmobilers meet others on the trail," National Trails Training Partnership website (October 1997): http://www.americantrails.org/resources/ManageMaintain/snowmeet.html (accessed April 29, 2015).
10. "Environmental Impacts from Snowmobile Use," Winter Wildlands Alliance (May 2014): http://winterwildlands.org/wp-content/uploads/2014/05/Environmental-Impacts-from-Snowmobile-Use.pdf (accessed April 29, 2015).
11. George Wuerthner, ed., *The Environmental Consequences of Motorized Recreation* (San Francisco, CA: The Foundation for Deep Ecology, 2008): accessed at http://www.stop-thrillcraft.org/kind_snowmobiles.htm (accessed April 29, 2015).
12. "Environmental Impacts from Snowmobile Use," Winter Wildlands Alliance.
13. Wuerthner, *The Environmental Consequences of Motorized Recreation*.
14. "Effects of Snowmobiling on Wildlife," *International Snowmobile Manufacturers' Association, Snowmobiling Fact Book*, http://www.snowmobile.org/facts_ece.asp (accessed April 29, 2015).

Chapter 38

1. "History of the Vardo (Gypsy Caravan)," GypsyWaggons.co.uk, http://gypsywaggons.co.uk/varhistory.htm (accessed June 23, 2014).
2. "The History of the RV in Canada and the U.S.," RV Hotline, http://www.rvhotlinecanada.com/rvhistory.asp (accessed June 23, 2014).
3. Julian Gothard, "The World's Ten Best Car Carrying Motorhomes," Examiner.com (August 26, 2012): http://www.examiner.com/article/the-world-s-ten-best-car-carrying-motorhomes (accessed June 24, 2014).

Chapter 39

1. J. Gordon Leishman, *Principles of Helicopter Aerodynamics, Cambridge aerospace series*, (Cambridge: Cambridge University Press, 2006), 18 (web extract).
2. Crouch, *Wings: A History of Aviation*, 26.
3. "Helicopter (Aerial Screw)," Leonardo da Vinci Inventions website, www.da-vinci-inventions.com/aerial-screw.aspx (accessed April 12, 2015).
4. Pendergast, *The Epic of Flight*, 73.
5. Pendergast, *The Epic of Flight*, 118.
6. Jeff Wise, "Finally! A Human-Powered Helicopter Wins the $250,000 Sikorsky Prize," *Popular Mechanics* (July 11, 2013): http://www.popularmechanics.com/flight/how-to/

a9138/finally-a-human-powered-helicopter-wins-the-250000-sikorsky-prize-15682369/ (accessed May 5, 2015).

7. Wise, "Finally!"

8. Adam Withnall, "Leonardo Da Vinci's dream realised: Canadian inventors achieve world's first man-powered helicopter flight," *The Independent* (July 12, 2013): http://www.independent.co.uk/news/world/americas/leonardo-da-vincis-dream-realised-canadian-inventors-achieve-worlds-first-manpowered-helicopter-flight-8706413.html (accessed May 5, 2015).

9. "Crane worker saved from Kingston fire in 'extremely unusual rescue,'" CBC News (last updated December 18, 2013): http://www.cbc.ca/news/canada/ottawa/crane-worker-saved-from-kingston-fire-in-extremely-unusual-rescue-1.2467711 (accessed May 5, 2015).

Chapter 40

1. Lao Long, "Rickshaw," Newsfinder (last updated June 2, 2003): http://www.newsfinder.org/site/more/rickshaw/ (accessed November 12, 2014).

2. Long, "Rickshaw."

3. Phillip Dixon, "Pedicabs Endure a Rough Ride," *Bicycle Times Magazine* (December 29, 2009): http://bicycletimesmag.com/pedicabs-endure-a-rough-ride/ (accessed November 14, 2014).

4. Anne Le Tran, "Rickshaws, Elvis, and Outside Lands: The History of Pedicabs," Esurance blog (posted July 24, 2012): http://blog.esurance.com/history-of-pedicabs/#.VFDP-cl50xjo (accessed November 11, 2014).

5. Jaya Jiwatram, "India's Cycle-Rickshaws Get a 'Solar' Upgrade: Indian research center unveils solar electric rickshaws to ease the country's traffic congestion and pollution woes," *Popular Science* (blog posted October 14, 2008): http://www.popsci.com/jaya-jiwatram/article/2008-10/indias-cycle-rickshaws-get-solar-upgrade (accessed November 12, 2014).

6. Helen Burchell, "Chen Guanming: The Olympic Chinese rickshaw rider's farewell to England," BBC News (August 23, 2013): http://www.bbc.com/news/uk-england-23792836 (accessed May 5, 2015).

7. Jeffrey E. Singer, "This Guy Who Spends His Life Riding Around the World in a Rickshaw Knows What's Up," Gothamist.com (September 2, 2015), http://gothamist.com/2015/09/02/rickshaw_hero_living_dream.php (accessed February 17, 2016).

Chapter 41

1. Natalie Rompella, *Famous Firsts: The Trendsetters, Groundbreakers & Risk-Takers Who Got America Moving!* (Montreal, Quebec: Lobster Press, 2007), 30, accessed at http://books.google.com/books?id=bc-_uWnoZWwC&pg=PA30#v=onepage&q&f=false (accessed July 20, 2014).

2. Steve Cave, "A Brief History of Skateboarding: Some big names and turning points in skate history," About.com, http://skateboard.about.com/cs/boardscience/a/brief_history.htm (accessed July 20, 2014).

3. Daniel Bradley, "Skateboards: Fit & Types," Livestrong.com (last updated October 30, 2013): http://www.livestrong.com/article/339215-skateboards-fit-types/ (accessed July 20, 2014).

4. Bradley, "Skateboards: Fit & Types."

5. Robert Behre, "Skateboarders face legal roadblocks," *The Post and Courier*, Charleston, SC (April 30, 2011): http://www.postandcourier.com/article/20110430/PC1602/304309937 (accessed July 20, 2014).

6. "Go Skateboarding Day," IASC, http://theiasc.org/go-skateboarding-day/ (accessed July 20, 2014).

7. "Stilts," Wikipedia, http://en.wikipedia.org/wiki/Stilts (accessed May 15, 2015).
8. "Roller Skates," Wikipedia, http://en.wikipedia.org/wiki/Roller_skates (accessed May 15, 2015).
9. "Pogo Stick," Wikipedia, http://en.wikipedia.org/wiki/Pogo_stick (accessed May 15, 2015).

Chapter 42

1. Sharolyn Herring and Christopher Fitzgerald, "History of the Hovercraft," Neoteric Hovercraft, Inc., http://www.neoterichovercraft.com/general_info/historyof.htm (accessed October 21, 2014).
2. "History of the Hovercraft," Discover Hover, http://www.discoverhover.org/abouthovercraft/history.htm (accessed October 21, 2014).
3. Herring and Fitzgerald, "History of the Hovercraft."
4. "Hampshire & Isle of Wight, Hovercraft History," BBC Home, http://www.bbc.co.uk/hampshire/content/articles/2008/03/27/history_hovercraft_feature.shtml (accessed October 21, 2014).
5. Herring and Fitzgerald, "History of the Hovercraft."
6. Herring and Fitzgerald, "History of the Hovercraft."

Chapter 43

1. Mary Bellis, "History of Electric Vehicles – The Early Years, Electric Cars from 1830 to 1930," About.com, http://inventors.about.com/od/estartinventions/a/History-Of-Electric-Vehicles.htm (accessed October 2, 2014).
2. Bellis, "History of Electric Vehicles."
3. Brad Berman, "History of Hybrid Vehicles," Hybridcars.com (June 14, 2011): http://www.hybridcars.com/history-of-hybrid-vehicles/ (accessed October 2, 2014).
4. Mark Rogowsky, "6 Reasons Tesla's Battery Swapping Could Take It to a 'Better Place,'" *Forbes Magazine* (June 21, 2013): http://www.forbes.com/sites/markrogowsky/2013/06/21/6-reasons-teslas-battery-swapping-could-take-it-to-a-better-place/ (accessed October 2, 2014).
5. Matt Lake, "How It Works; A Tale of 2 Engines: How Hybrid Cars Tame Emissions," *New York Times, Technology* (November 8, 2001): http://www.nytimes.com/2001/11/08/technology/how-it-works-a-tale-of-2-engines-how-hybrid-cars-tame-emissions.html?scp=1&sq=hybrid%20Toyota%20Prius%201997%20Honda%20Insight%201999&st=cse (accessed October 3, 2014).
6. Ozzie Zehner, "Unclean at Any Speed: Electric cars don't solve the automobile's environmental problems, "IEEE Spectrum (blog posted June 30, 2013): http://spectrum.ieee.org/energy/renewables/unclean-at-any-speed (accessed October 3, 2014).
7. Angela Greiling Keane, "US will require too-quiet electric cars to make noise, Rule would apply under 18 m.p.h," *Boston Globe* (January 08, 2013): http://www.bostonglobe.com/business/2013/01/08/government-require-electric-cars-make-noise/bGmucrhawg3ftkIi-7HqdQO/story.html (accessed October 3, 2014).

Chapter 44

1. "Facts and History about ATV," ATV Quad News, http://www.atv.info/page.cfm?name=ATV%20Facts (accessed October 24, 2014).
2. "The History of the ATV," OutdoorHub, http://www.outdoorhub.com/stories/2012/08/01/the-history-of-the-atv/ (accessed October 24, 2014).
3, "Facts and History about ATV."
4. "The History of the ATV."
5. "Facts and History about ATV."

6. V. Bonsal, et al., "A 21-year history of all-terrain vehicle injuries: has anything changed?" *American Journal of Surgery* (June 2008): 789–792 abstracted at http://www.ncbi.nlm.nih.gov/pubmed/18367134?report=abstract (accessed October 24, 2014).

7. "The Environmental Impact of ATVs," Portland Natural Gas Transmission System, http://www.pngts.com/atv.html (accessed April 30, 2015).

Chapter 45

1. Grant, *Flight, The Complete History*, 334.
2. Crouch, *Wings: A History of Aviation*, 196.
3. Grant, *Flight, The Complete History*, 340–341.
4. Grant, *Flight, The Complete History*, 342.
5. Grant, *Flight, The Complete History*, 344.
6. Grant, *Flight, The Complete History*, 350.
7. Grant, *Flight, The Complete History*, 373.
8. Robert A. Braeunig, "Did We Land on The Moon? A Debunking of the Moon Hoax Theory," http://braeunig.us/space/hoax.htm (accessed May 6, 2015).
9. "Remotely Operated Underwater Vehicle," Wikipedia, https://en.wikipedia.org/wiki/Remotely_operated_underwater_vehicle (accessed December 28, 2015).
10. "Exploration of Mars," Wikipedia, https://en.wikipedia.org/wiki/Exploration_of_Mars (accessed December 28, 2015).

Chapter 46

1. Thomas Ricker, "The Future Passed: Jetpack Edition," *The Verge* (November 3, 2011): http://www.theverge.com/2011/11/3/2504531/jetpack-history-future-passed (accessed November 4, 2014).
2. Ricker, "The Future Passed."
3. Ricker, "The Future Passed."
4. Ricker, "The Future Passed."
5. Ricker, "The Future Passed."
6. "The Martin Jetpack – Fly the Dream," Martin Aircraft Company, http://www.martinjetpack.com/ (accessed November 4, 2014).
7. Ricker, "The Future Passed."
8. "Jetman Dubai – The Next Chapter: Yves Rossy & Vince Reffet," Jetman, http://www.jetman.com/?page_id=24 (accessed November 4, 2014).

Chapter 47

1. Frew, *Locomotives*, 115.
2. Frew, *Locomotives*, 120.
3. Frew, *Locomotives*, 122.
4. Frew, *Locomotives*, 132.
5. Frew, *Locomotives*, 136.
6. "Shinkansen," Japan-guide.com, http://www.japan-guide.com/e/e2018.html (accessed September 9, 2014).
7. "China High-Speed Train (Bullet Train)," TravelChinaGuide.com, http://www.travelchinaguide.com/china-trains/high-speed/ (accessed September 9, 2014).
8. *High-Speed Europe: A Sustainable Link Between Citizens* (Luxembourg: Publications Office of the European Union, 2010): http://ec.europa.eu/transport/themes/infrastructure/studies/doc/2010_high_speed_rail_en.pdf (accessed September 9, 2014).
9. "The Development of High Speed Rail in the United States: Issues and Recent Events," U.S. Congressional Research Service (December 20, 2013): http://fas.org/sgp/crs/misc/R42584.pdf (accessed September 9, 2014).

10. "How Maglev Trains Work," How Stuff Works, http://science.howstuffworks.com/transport/engines-equipment/maglev-train.htm (accessed September 9, 2014).
11. "How Maglev Trains Work."
12. "New maglev Shinkansen to run underground for 86% of initial route," *The Ashai Shimbun* (September 19, 2013): http://ajw.asahi.com/article/behind_news/social_affairs/AJ201309190078 (accessed September 9, 2014).
13. Alan Kandel, "Conventional High-Speed Rail vs. Magnetically Levitated Trains: Was Maglev Ever In Contention?" California Progress Report (blog posted November 22, 2011): http://www.californiaprogressreport.com/site/conventional-high-speed-rail-vs-magnetically-levitated-trains-was-maglev-ever-contention (accessed September 10, 2014).
14. Laurence E. Blow, "Dispelling the Top Ten Myths of Maglev," *High Speed Rail 2010 Conference White Paper* accessed at: https://faculty.washington.edu/jbs/itrans/dispelling-myths-blow.pdf (accessed September 9, 2014).

Chapter 48
1. Crouch, *Wings: A History of Aviation*, 455–458.
2. Crouch, *Wings: A History of Aviation*, 463.
3. Crouch, *Wings: A History of Aviation*, 526.
4. Grant, *Flight, The Complete History*, 393.
5. Crouch, *Wings: A History of Aviation*, 629.

Chapter 49
1. Grant, *Flight, The Complete History*, 340–341.
2. Crouch, *Wings: A History of Aviation*, 463.
3. Grant, *Flight, The Complete History*, 360.
4. Grant, *Flight, The Complete History*, 363.
5. "The Shuttle," NASA, http://www.nasa.gov/externalflash/the_shuttle/ (accessed May 6, 2015).
6. Grant, *Flight, The Complete History*, 436.
7. Jonathan Amos, "Richard Branson unveils Virgin Galactic spaceplane," BBC News (December 8, 2009): http://news.bbc.co.uk/2/hi/science/nature/8400353.stm (accessed May 5, 2015).
8. Kenneth Chang and John Schwartz, "Virgin Galactic's SpaceShipTwo Crashes in New Setback for Commercial Spaceflight," *The New York Times* (October 31, 2014): http://www.nytimes.com/2014/11/01/science/virgin-galactics-spaceshiptwo-crashes-during-test-flight.html?_r=0 (accessed May 5, 2015).
9. Chang and Schwartz, "Virgin Galactic's SpaceShipTwo Crashes."
10. Chang and Schwartz, "Virgin Galactic's SpaceShipTwo Crashes."
11. Ryan Whitwam, "SpaceX says Falcon 9 rocket is undamaged after historic landing," *ExtremeTech* (January 4, 2016): http://www.extremetech.com/extreme/220276-spacex-says-falcon-9-rocket-is-undamaged-after-historic-landing (accessed January 6, 2016).

Chapter 50
1. Tom Harris, "How Segways Work," How Stuff Works, http://science.howstuffworks.com/engineering/civil/ginger.htm (accessed June 4, 2014).
2. Segway Inc., http://www.segway.com/ (accessed June 2, 2014).
3. Segway Inc. website
4. "Segway Laws," Governors Highway Safety Association, http://www.ghsa.org/html/stateinfo/laws/segway_laws.html (accessed June 4, 2014).
5. Fran Spielman, "City puts brake on Segway tours," *Chicago Sun Times* (May 28, 2014): https://web.archive.org/web/20151003211130/http://chicago.suntimes.com/?p=165263 (accessed April 2, 2016).

6. "Segway Polo," Wikipedia, http://en.wikipedia.org/wiki/Segway_polo (accessed June 4, 2014).

7. "Segway, Inc." Wikipedia, http://en.wikipedia.org/wiki/Segway_Inc (accessed June 4, 2014).

Conclusion

1. Eileen Gunn, "How America's Leading Science Fiction Authors Are Shaping Your Future," *Smithsonian Magazine* (May 2014): http://www.smithsonianmag.com/arts-culture/how-americas-leading-science-fiction-authors-are-shaping-your-future-180951169/?no-ist (accessed January 6, 2016).

2. Gunn, "How America's Leading Science Fiction Authors Are Shaping Your Future."

3. John Hutchinson, "The future of travel revealed…from vertical plane seats to 800mph trains…and a virtual Stonehenge," *Daily Mail* (or MailOnline?) (October 25, 2015): http://www.dailymail.co.uk/travel/travel_news/article-3262149/Planes-transparent-walls-800mph-trains-vacuum-tunnels-computers-tell-destinations-like-future-travel-revealed.html (accessed January 9, 2016).

4. Will Coldwell, "The future of travel: what will holidays look like in 2024?" *The Guardian*, September 29, 2014.

Index

Author and Series Presenter Biographies

PAULA GREY is a writer of short fiction and non-fiction and a former grant writer, software technical writer, and technical editor. This is her first book. Paula and her husband live in Wakefield, Rhode Island, and Paula enjoys traveling by land, sea, or air whenever she gets the opportunity.

PHILLIP HOOSE is the widely acclaimed author of books, essays, stories, songs, and articles, including the National Book Award and Newbery Honor winning book *Claudette Colvin: Twice toward Justice* and the Boston Globe–Horn Book Honor winner *The Boys Who Challenged Hitler: Knud Pedersen and the Churchill Club.* A graduate of Indiana University and the Yale School of Forestry and Environmental Sciences, Hoose was for 37 years a staff member of The Nature Conservancy, dedicated to preserving the plants, animals, and natural communities of the Earth. Find out more at www.philliphoose. com. (Photo by Gordon Chibrowski, Maine Newspapers)